Border Crossings

The border between intimate memory and historical revelation is explored in this wide-ranging collection, which features original contributions from leading figures in the life-writing field from Australia, Canada, Europe, the UK, and the USA.

The transmission and preservation of personal knowledge and stories from generation to generation frequently requires crossing into the private, contested spaces of memory. The most secret accounts or guarded remnants of information can sometimes lead to the most profound insights. In this context, there is a delicate balance between life writing's role in revealing lives and the desire to be respectful towards them. As the essays in this book attest, exposing secrets, even if humiliating, can be a way of honouring lives. Throughout runs the framing theme of memory as the source of all intergenerational transmission of culture and history—whether relating to family, community, nation, ancestry, or political allegiance—and the importance of the intimate and personal in that process of handing on.

This book was originally published as a special issue of *Life Writing*.

Paul Longley Arthur is Professor in Digital Humanities at Western Sydney University, Australia. He was previously Deputy Director of the National Centre of Biography and the Centre for European Studies at the Australian National University and Deputy General Editor of the *Australian Dictionary of Biography*. He has published widely in cultural and communication studies, history, literature, and media.

Leena Kurvet-Käosaar is Associate Professor of Literary Theory, and Programme Director of Literature and Cultural Research, at the University of Tartu, Estonia, as well as a Senior Researcher at the Estonian Literary Museum. She has published widely in life-writing studies, specialising in particular on Post-Soviet life writing, personal narratives of Soviet deportations and the Gulag, and trauma studies.

Life Writing
Academic Editor: Maureen Perkins, Macquarie University, Australia

Life Writing, founded in 2004 by Mary Besemeres and Maureen Perkins, is one of the leading journals in the field of biography and autobiography.

Its title indicates that it reaches beyond traditional interpretations of biography and autobiography as genres belonging solely in the study of literature. It welcomes work from any discipline that discusses the nature of the self and self-expression and how these interact with the process of recording a life. Life writing is about expanding the ways in which we understand how lives are represented.

The journal has a special, though not exclusive, interest in cross-cultural experience. It also has the unique and unusual policy of carrying both scholarly articles and critically informed personal narratives. It is published four times a year, and its editorial board comprises leaders in the field of life-writing practice.

Book titles from *Life Writing* include:

Trauma Texts
Edited by Gillian Whitlock and Kate Douglas

Poetry and Autobiography
Edited by Jo Gill and Melanie Waters

International Life Writing
Memory and Identity in Global Context
Edited by Paul Longley Arthur

Dissenting Lives
Edited by Anne Collett and Tony Simoes da Silva

Border Crossings
Essays in Identity and Belonging
Edited by Paul Longley Arthur and Leena Kurvet-Käosaar

Border Crossings
Essays in Identity and Belonging

Edited by
**Paul Longley Arthur and
Leena Kurvet-Käosaar**

LONDON AND NEW YORK

First published 2019
by Routledge
2 Park Square, Milton Park, Abingdon, Oxon, OX14 4RN, UK

and by Routledge
711 Third Avenue, New York, NY 10017, USA

Routledge is an imprint of the Taylor & Francis Group, an informa business

© 2019 Taylor & Francis

All rights reserved. No part of this book may be reprinted or reproduced or utilised in any form or by any electronic, mechanical, or other means, now known or hereafter invented, including photocopying and recording, or in any information storage or retrieval system, without permission in writing from the publishers.

Trademark notice: Product or corporate names may be trademarks or registered trademarks, and are used only for identification and explanation without intent to infringe.

British Library Cataloguing-in-Publication Data
A catalogue record for this book is available from the British Library

ISBN13: 978-1-138-67109-6

Typeset in Myriad Pro
by codeMantra

Publisher's Note
The publisher accepts responsibility for any inconsistencies that may have arisen during the conversion of this book from journal articles to book chapters, namely the possible inclusion of journal terminology.

Disclaimer
Every effort has been made to contact copyright holders for their permission to reprint material in this book. The publishers would be grateful to hear from any copyright holder who is not here acknowledged and will undertake to rectify any errors or omissions in future editions of this book.

Contents

Citation Information   vi
Notes on Contributors   viii

Introduction: Thresholds and Boundaries   1
Paul Longley Arthur and Leena Kurvet-Käosaar

1. Family Memoir and Self-Discovery   7
Jeremy D. Popkin

2. Voices in Movement: Feminist Family Stories in Oral History and Sound Art   19
Margaretta Jolly

3. The Epistolary Dynamics of Sisterhood Across the Iron Curtain   41
Leena Kurvet-Käosaar

4. The Odyssey Quilts: Narrative Artworks of Childhood, War, and Migration   57
Nonja Peters

5. Is Autobiographical Writing a Historical Document?: The Impact of Self-Censorship on Life Narratives   97
Magda Stroińska and Vikki Cecchetto

6. Material Memory and the Digital   109
Paul Longley Arthur

7. Because it's Your Country: Death and its Meanings in West Arnhem Land   123
Martin Thomas

8. 'from Organic Acts': Tsamorita, Rosaries, and the Poem of My Grandma's Life   145
Craig Santos Perez

Index   151

Citation Information

The chapters in this book were originally published in *Life Writing*, volume 12, issue 2 (June 2015). When citing this material, please use the original page numbering for each article, as follows:

Chapter 1
Family Memoir and Self-Discovery
Jeremy D. Popkin
Life Writing, volume 12, issue 2 (June 2015) pp. 127–138

Chapter 2
Voices in Movement: Feminist Family Stories in Oral History and Sound Art
Margaretta Jolly
Life Writing, volume 12, issue 2 (June 2015) pp. 139–160

Chapter 3
The Epistolary Dynamics of Sisterhood Across the Iron Curtain
Leena Kurvet-Käosaar
Life Writing, volume 12, issue 2 (June 2015) pp. 161–176

Chapter 5
Is Autobiographical Writing a Historical Document?: The Impact of Self-Censorship on Life Narratives
Magda Stroińska and Vikki Cecchetto
Life Writing, volume 12, issue 2 (June 2015) pp. 177–188

Chapter 6
Material Memory and the Digital
Paul Longley Arthur
Life Writing, volume 12, issue 2 (June 2015) pp. 189–200

Chapter 7
Because it's Your Country: Death and its Meanings in West Arnhem Land
Martin Thomas
Life Writing, volume 12, issue 2 (June 2015) pp. 203–224

Chapter 8

'from *Organic Acts*': Tsamorita, Rosaries, and the Poem of My Grandma's Life
Craig Santos Perez
Life Writing, volume 12, issue 2 (June 2015) pp. 225–230

For any permission-related enquiries please visit:
http://www.tandfonline.com/page/help/permissions

Notes on Contributors

Paul Longley Arthur is a Professorial Research Fellow in the School of Arts and Humanities at Edith Cowan University, Australia. He has published widely in cultural and communication studies, biography, history and literature, and is the author of *Virtual Voyages: Travel Writing and the Antipodes, 1605–1837* (2010). Recent edited volumes include *International Life Writing: Memory and Identity in Global Context* (2013), *Framing Lives* (2014), and *Migrant Nation: Australian Culture, Society and Identity* (2018).

Vikki Cecchetto (now retired as Associate Professor of Italian and Linguistics at McMaster University, Hamilton, Canada), continues her research focused now primarily on second-language attrition in aging immigrants, communication with bilinguals with dementias, and autobiographical / trauma narratives. She is the editor of *Exile, Language and Identity* (with Stroińska, 2003), *International Classroom: Challenging the Notion* (with Stroińska, 2006), and *The Unspeakable: Narratives of Trauma* (with Stroinska and Szymanski, 2014).

Margaretta Jolly is Reader in Cultural Studies at the University of Sussex, UK, and directs the University's Centre for Life History and Life Writing Research. Her work has focused on auto/biography, letter writing, and oral history, particularly in relation to women's movements. Her book *In Love and Struggle* won the Feminist Women's Studies Association UK prize in 2009. She directed Sisterhood and After: The Women's Liberation Oral History Project with The British Library from 2010–2013 and now leads The Business of Women's Words: Purpose and Profit in Feminist Publishing, also partnered with The British Library.

Leena Kurvet-Käosaar is Associate Professor of Cultural Theory at the University of Tartu, Estonia, as well as a Senior Researcher at the Estonian Literary Museum and also the leader of the research group on migration and diaspora studies of the Center of Excellence of Estonian Studies at the Estonian Literary Museum. She has published widely in life-writing studies, specialising in particular on Post-Soviet life writing, personal narratives of Soviet deportations and the Gulag, and trauma studies.

Craig Santos Perez is an Associate Professor in the English department at the University of Hawai'i, Mānoa. He is the author of four books of poetry and the co-editor of three anthologies of Pacific Islander literature.

Nonja Peters is an historian, anthropologist, museum curator, and social researcher whose expertise is transnational migration (forced and voluntary) and resettlement in Australia: ethnicity, sense of place, identity and belonging, immigrant entrepreneurship, racism, and the sustainable preservation of immigrants' tangible, intangible, and digital cultural heritage. She has a special interest in Dutch maritime, military, migration, and mercantile connections with Australia and the South East Asian Region since 1606. She is currently involved in academic, community-based, visual, and bilateral research, and events in all these areas in Australia and internationally.

Jeremy D. Popkin is the William T. Bryan Chair of History at the University of Kentucky USA. He is the author of *History, Historians and Autobiography* (2005) and co-editor, with Julie Rak, of *On Diary* (2009), a collection of Philippe Lejeune's essays on personal journals. He has also published extensively on the history of the French and Haitian Revolutions.

Magda Stroińska is Professor of German and Linguistics at McMaster University in Hamilton, Ontario, Canada. Her major area of research is cognitive sociolinguistics, in particular propaganda and language manipulation, issues of identity in exile, aging, and bilingualism, as well as language and psychological trauma. She currently studies hate speech in totalitarian regimes and its effects in post-communist Eastern Europe and beyond. She has edited a number of books, including volumes on linguistic representations of culture (2001), on exile, language, and identity (with Vikki Cecchetto, 2003), and on trauma narratives (with Cecchetto and Szymanski, 2014).

Martin Thomas is a cultural historian who specialises in Australian, Aboriginal, and trans-national history. He has published in the areas of environmental history, landscape studies, cross-cultural encounter, expeditions and exploration, history of anthropology, and on the impact of sound recording and photography. His recent teaching at ANU includes courses on exploration, photography, and public history. The film described in this essay, titled *Etched in Bone* (2018), is distributed by Ronin Films.

Introduction: Thresholds and Boundaries

Paul Longley Arthur and Leena Kurvet-Käosaar

Running through the essays in this collection is the major theme of memory as the source of intergenerational transmission of culture and history across borders of many kinds—whether relating to family, community, nation, ancestry, or political allegiance—and the importance of the intimate and personal in that transmission. In any attempt to report on the life of another, or even one's own life, an inescapable ethical dilemma arises that relates to entering intensely private areas of experience and presenting an intimate subject matter for the world to see. How much intimate material should be revealed? For what purpose? To whose benefit? At what risk? How? In an era when millions of people are willing to share the minutiae of their individual daily lives via social media and the private lives of the famous are exposed routinely to mass audiences, such questions loom larger than ever. With easier access to private information—by governments, hackers, marketers, and private citizens—this area has become one of global concern in the context of the fundamental human right to privacy.[1]

The transmission and preservation of stories from generation to generation frequently require crossing into private spaces.[2] Critical engagement with the private and the intimate has always been a key characteristic of life-writing studies, a field which has made a noteworthy contribution to contemporary re-conceptualisations of the private and the public spheres and the intricate interconnections between them. Life writing frequently needs to use imaginative and fictional strategies to overcome gaps and absences while importantly retaining an awareness of the transformations and substitutions taking place in the course of their utilisation (Miller; Hoffmann). Life writing is also required to perform acts of interpretation and translation—in the figurative and literal sense—concerning, for example, intergenerational acts of transfer (Hirsch) that may involve crossing languages, cultural contexts, time periods, or political ideologies. Further, there can be discord between place as a geographical entity today and its memorial implications with regard to lost and destroyed realities (Hirsch and Spitzer). For many life-writing scholars, their own family history has constituted a central site of exploration that has informed and shaped their theoretical perceptions. When dealing with intimate material, the choice of style, media, and degree of imaginative intervention can be a sensitive ethical as well as aesthetic matter.

Sometimes the most secret memories or the most private remnants of information are also those that can lead to the most profound insights. Exposing secrets, even humiliating secrets, can then become a way of honouring lives. But how much personal information about a subject, often revealed posthumously via documents and artefacts left behind, can be exposed without threatening or invading personal privacy? The delicate balance between life writing's need to delve into the lives of others and the desire to be respectful towards them is the topic that each of the essays in this collection confronts, explores, or embodies. This, in turn, highlights an awareness of the relational nature of privacy as well as the need to take into consideration the specific characteristics of the type of medium or genre involved: as Margaretta Jolly has formulated it, 'understanding and respecting the kind of sociability each form of address presupposes' (232). These concerns also highlight anew the ways in which forms of autobiographical truth are tightly interwoven with issues of memory, representation, and ethics.

There are many forms of auto/biographical truth. In his seminal 'The Autobiographical Pact' Philippe Lejeune formulates the notion of 'the referential pact' as a 'supplementary proof of honesty' that outlines the degree of 'the possible', which functions 'to indicate explicitly the *field* to which this oath [to tell the truth] applies' (Lejeune 22). Elizabeth Bruss emphasises the need for an autobiographical act to make two kinds of claims for truth: first, 'a claim ... for the truth-value of what the autobiography reports', and second, to establish that the autobiographer 'purports to believe in what he asserts' (Bruss 11). While Lejeune and Bruss acknowledge the difficulty of verifying truth claims, they both see these forms of personal truth as fundamental to autobiography. For Sidonie Smith and Julia Watson the self-referentiality that is a core trait of autobiography means that 'the truth of the [autobiographical] narrative ... can be neither fully verified nor fully discredited'. Rather the truth is the kind that 'resides in the intersubjective exchange between narrator and reader ... produc[ing] a polyphonic site of indeterminacy rather than a single, stable truth' (Smith and Watson 16). Such 'communicative action' between the narrator and addressee(s) plays a key role in the constitution of an autobiographical act and the emergence of 'intersubjective truths that are constructed in this process' (Smith and Watson 90).

The two paradigms of autobiographical truth, though they may seem to be mutually exclusive, have recently (re)emerged both in life-writing practices and life-writing scholarship in ways that forge ties between an assessment of the factual and verifiable foundation of life writing and the dialogic, intersubjective, and polyphonic nature of auto/biographical truth(s). As several essays in the current collection demonstrate, a shift can be traced from sole reliance on the power of memory (intertwined with the power of imagination in making up for the gaps in memory) to engagement with factual, documentary evidence of lives. Such evidence is not attributed a definite truth value that outweighs the role of memory and imagination; rather, the reader is provided access to the complex processes in which such documents come to acquire meaning in life writing. These include processes of verification and contextualisation that make visible more general, external, historical, and cultural implications, as well as personal and private perspectives that shape the representation of life experience from within (Kraus 245, Miller 4–5).

The ways in which auto/biographical truths are presented as intersubjective and polyphonic bring into focus not only the potential of address and exchange between the author or the one who represents lives and the reader but just as importantly how individual lives are always inextricably related to those of others—for example, through family ties and affiliations, modes of communal identification, cultural and historical contexts and legacies, trajectories of the travel of memory, and connections to geographical place. All contributions in the volume underline the need to carefully attend to the ways in which these ties and connections are made visible in the process of narrating lives and the implications of different poetic and imaginative modes of self-expression or storytelling and/or research trajectories. Furthermore, what is highlighted here is not only a recognition that a representation of one's life inevitably concerns the lives of others but that such relational self-perception is often at the heart of an auto/biographical quest. Therefore, it can be said that the conceptualisations of the relational nature of life writing, first articulated by scholars of women's autobiography, attest with new vigour to its applicability to the process of narrating lives on a more general level.

The sense of intergenerational obligation in these essays adds urgency to the task of preservation while foregrounding the special kinds of courtesies required when the subjects are from prior generations, including the distant past. Permission to publish can be sought from the living but not the dead. A related matter is the practical task of locating clues and traces and finding appropriate ways to use them—whether they are material items, such as letters, diaries, or photographs, or they take the form of spoken words, in which case the voice, language, and speech patterns become an important part of the subject's portrayal and the world that person inhabited. In intimate exchanges, silences and hesitations can be as significant as the words themselves. These 'clues' do not speak for themselves. In order for them to make sense they need to be followed up and researched—placed into historical, cultural, and geographical contexts. New knowledge emerges in the process, but this also bears important ethical implications. Taking the time and energy to seek and verify that evidence, to put it into proper cultural-historical contexts, points to an ethical obligation towards the life of another. Whether it concerns family history or a larger cultural or historical phenomenon, reliance on memory, imagination, and artistic perspective is not sufficient. Documentary research is an ethical imperative that needs to be carried out for an accountable and respectful representation of someone else's life.

Yet documentary or verifiable evidence alone hardly ever constitutes the definitive resource for writing and researching lives: memories remain a crucial and essential part of the process of discovering, uncovering, and honouring lives. Essays in this collection focus on a diverse range of aspects of memory, concerning its modes of mediation and transmission, its intersubjective nature, its representational possibilities and limits, as well as its relationship with other kinds of resources for writing and researching lives. Memory can be experienced and transmitted in many different ways, and it may be, as Paul Arthur's essay in this volume suggests, that processes of remembering, as well as methods of storing and sharing memories, are currently being influenced and changed by digital technologies. There may be a need to develop strategies for forgetting as every life amasses uncontrollable quantities of data on multiple devices and on the Web. In this context, the sharing of life experiences and memories through letters, notebooks, or face-to-face contact is celebrated in this collection as something to be treasured.

Craig Santos Perez begins his essay by reproducing his grandmother's words, spoken in the Chamorro language, just as he remembers hearing them as a child. 'Her voice and the words of our native language', he explains, 'form a canoe that carries me on the currents of memory back to our home island of Guåhan (Guam) in the western Pacific Ocean'. Inserting her words into the poetry he writes and publishes allows Perez to tell his grandmother's story as part of his own. At the same time, his essay tells the story of his country's experience of colonisation, with the quoted fragments of his grandmother's speech bringing the now-endangered Chamorro language of his childhood into the present, celebrating it and connecting his life with hers in a way that expresses loss as well as continuing connection: 'Poetry weaves our voices together' (in this volume).

For Jeremy D. Popkin the painstaking process of searching for and compiling material for a family memoir inevitably also reflects back upon the searcher's own life: 'The product of all this effort is both a biography or a history, and also a kind of autobiography' (in this volume). With first-hand experience through his own family history project, of the hazards of entering private territory, Popkin offers a critical overview of a selection of family memoirs that have dealt in various ways with an intimate intergenerational subject matter. If any intimate material is to be revealed, he argues, verifiable documentation and factual accuracy are essential:

> In an age all too prone to assume that memory, with its emotional appeal and its link to fiction, is more important to human life than the factual record, projects like these reassert the value of a more critical view, even of the personal past.
> (in this volume)

Drawing upon a rich family archive of correspondence between two Estonian sisters, separated for many years by the Iron Curtain but reunited through their letters, Leena Kurvet-Käosaar's essay shows how factual accuracy and truth-telling are not possible in some circumstances. In order to elude political censorship, the sisters, one of whom is the author's grandmother, develop narrative strategies that amount to private codes for communicating 'facts' about post-war life in Estonia without explicitly revealing them. However, despite censorial concerns and lives lived within vastly different socio-political contexts with minimal possibility for face-to-face encounters, the correspondence offers proof of the capacity of the epistolary medium to facilitate a long-lasting and intimate sisterly bond.

Nonja Peters's account of the Dutch Odyssey Quilts, a set of three wall hangings that portray their creators' memories of childhood, the Second World War, and migration to Australia, demonstrates that these ornate visual narratives act as a vital, recuperative story-form. Now part of the permanent collection of the Sydney Power-House Museum, the quilts are the product of close collaboration between ten Dutch women over five years. They are made up of many individually designed and crafted quilting pieces, each of which tells a story. Read in conjunction with the written notes, as well as the mementos, sketches, and photographs that make up the artists' visual diaries, these remarkable works display the way that deeply personal and fragmentary recollections, sometimes of brief intense moments, can be pieced together and transformed to create a powerful collective history of war, migration, and displacement. The private has been made public in a way that enables the preservation and sharing of specific historical experiences that might otherwise be forgotten, at the same time as connecting these stories with the broader history of post-war migration to Australia.

In the context of life narratives, the factual record is always as open to processes of selection, interpretation, occlusion, or forgetting as any other, as Magda Stroińska and Vikki Cecchetto demonstrate in an essay that explores self-censorship in autobiographical works. Their focus is on the Polish writer Andrzej Czcibor-Piotrowski and, in particular, his two very different accounts, published almost four decades apart, of his childhood spent in deportation camps in Russia during the First World War. In the earlier work, externally imposed Soviet censorship explains his omission of key historical events and their traumatic effects. However, by the time he wrote the later work, there was no longer any such external pressure. In this case, the explanation the authors offer for Czcibor-Piotrowski's oblique and muted handling of the trauma he undoubtedly experienced as a child has relevance to trauma testimony more generally: '[T]he reported memories mask those that may be too painful to be revealed' (in this volume). For the 'wounded storyteller' the imperative to bury or forget the long-held intimate secrets of the past—or simply to shield one's private self from public scrutiny—can overwhelm the desire to share one's story.

Conversely, in many cases the subject accepts the opportunity to expose painful personal secrets. Margaretta Jolly reports on a remarkable oral history project that has transmuted recorded fragments of spoken memories of deeply private, traumatic experiences into an art installation for public consumption in a gallery setting. The two subjects are women from a larger group of interviewees for the recent 'Sisterhood and After: Women's Liberation Oral History Project' that captures 60 women's accounts of their political activism from the 1960s to the 1980s in the United Kingdom for archiving in the British Library. The two women brought together in this performance work each hold long-suppressed family secrets. While their paths have never crossed in life, their voices intertwine and weave through a soundscape of music and other sounds as memory fragments float out, accompanied by fleeting visual images from the past. As well as being the site of self-exposure, the installation becomes an effective medium for the exploration of changing concepts of family over the past half century and, in particular, the complex and unstable relationship between feminist and family values.

The essays in this collection demonstrate that there are many ways of dealing with the challenges of negotiating, in life writing, the unstable border between historical revelation and personal violation; the choices are driven by the specific circumstances of each life and each context, including the extent of potential hurt that can be inflicted and the nature of the relationship between the writer/reporter and the subject. While memories can inflict pain by opening old wounds, they can also be recuperative. In telling the story of the recent death of a friend—the Australian Indigenous leader known as Wamud—and sharing memories of his life and their friendship, Martin Thomas takes his readers on an extraordinary journey into Arnhem Land in the Northern Territory of Australia.[3] Within Thomas's narrative is an intensely moving account of the repatriation of the bones of Arnhem Land ancestors from the Smithsonian Museum in Washington, DC, where they had been deposited after a scientific expedition removed them from their burial place in 1948. With symbolic reverberations far beyond the life of Wamud and his ancestors, this story takes us back to a colonial past where anthropological intrusion was routinely sanctioned as scientific research. Sadly, it also confronts us with shameful contemporary realities relating to the lives of Indigenous people in Australia. Through its multiple generational and cultural lenses, and particularly through the story of the life and death of a contemporary elder, this narrative dramatizes a deeply paradoxical dilemma facing descendants and the wider community. In an ironic twist, it becomes clear that the products of past anthropological intrusions—notably the 1948 archival film footage recording a sacred initiation ceremony—while serving as a symbol of colonial trespass into forbidden territory, may at the same time offer a vital key to preserving and passing on knowledge of a cultural tradition built over many millennia, but ravaged and decimated in the space of less than two centuries.

Acknowledgement

This article was supported by the Estonian Ministry of Education and Research (IUT22-2) and by the European Union through the European Regional Development Fund (Centre of Excellence in Estonian Studies).

Notes

[1] See Report of the Office of the United Nations High Commissioner for Human Rights:

> In its resolution 68/167, the General Assembly requested the United Nations High Commissioner for Human Rights to submit a report on the protection and promotion of the right to privacy in the context of domestic and extraterritorial surveillance and/or the interception of digital communications and the collection of personal data. (1)

'As contemporary life is played out ever more online, the Internet has become both ubiquitous and increasingly intimate' (3).

[2] A number of the essays in this special issue originated as presentations given at the Private Lives, Intimate Readings symposium convened by Leena Kurvet-Käosaar at the University of Tartu, Estonia in 2013.
[3] Martin Thomas's essay, 'Because it's Your Country', won the Australian Book Review's Calibre Prize 2013 and was first published in the Australian Book Review, No. 350, April 2013, pp. 26–37.

References

Bruss, Elizabeth. *Autobiographical Acts: The Changing Situation of a Literary Genre*. Baltimore, MD: Johns Hopkins UP, 1976.
Hirsch, Marianne. 'The Generation of Postmemory'. *Poetics Today* 29.1 (2008): 103–28.
Hirsch, Marianne, and Leo Spitzer. *Ghosts of Home: The Afterlife of Czernowitz in Jewish Memory*. Berkeley: U of California P, 2010.
Hoffmann, Eva. *After Such Knowledge: Memory, History, and the Legacy of the Holocaust*. New York: Public Affairs, 2004.
Jolly, Margaretta. *In Love and Struggle: Letters in Contemporary Feminism*. New York: Columbia UP, 2008.
Lejeune, Philippe. 'The Autobiographical Pact.' *On Autobiography*. Ed. Palu J. Eakin. Minneapolis: U of Minnesota Press, 1989. 3–30.
Miller, Nancy. *What They Saved: Pieces of Jewish Past*. Lincoln. U of Nebraska P, 2011.
Smith, Sidone and JuliaWatson. *Reading Autobiography. A Guide for Interpreting Life Narratives*. Second Edition. Minneapolis: U of Minnesota Press, 2010.
United Nations, Office of the High Commissioner for Human Rights. *The Right to Privacy in the Digital Age*. Accessed 7 January 2015. http://www.ohchr.org/EN/Issues/DigitalAge/Pages/DigitalAgeIndex.aspx

Family Memoir and Self-Discovery

Jeremy D. Popkin

In the increasingly popular genre of 'family memoir,' authors take readers with them as they pursue the details of the lives of their parents and other relatives. Emphasising both the painstaking quest for information about a family past they did not know and the highly personal nature of such projects, family memoirs straddle the boundaries between history, biography, and autobiography. Family memoir is an aspect of the graphic-art works of Art Spiegelman and Alison Bechdel, but it has also been pursued by academics and journalists, including Daniel Mendelsohn, Alexander Stille, and Bliss Broyard. Often resented by their authors' living relatives, these chronicles appeal to readers because they pose universal questions about the connections between family history and personal identity.

What is a 'family memoir'? Like so many other labels in life-writing studies, the term is often used but rarely defined. In their survey of life writing, *Reading Autobiography,* Sidonie Smith and Julia Watson deal with 'narratives of family' in a section titled 'New-Model Narratives of Displacement, Migration, and Exile'. They see the phenomenon as a 'response to the cultural sense that 'the family' as an institution is under assault in the West because of migratory societies, smaller families, greater mobility, and women's increasing participation in work and public life', which has driven readers to write such stories for 'an audience seeking cohesion'. This might suggest narratives coloured by nostalgia for a lost world of stable extended families. The examples that Smith and Watson cite, however, are most often stories of dysfunctional or broken families, searches for missing parents, and discoveries of hidden secrets whose publication, the authors note, sometimes 'disturb[s] living parents or siblings' and thus provokes new forms of family discord (154–56).

The group of 'family memoirs' I discuss in this paper are neither nostalgic evocations of sheltering family Edens whose loss the authors lament nor dissections of notably dysfunctional households.[1] In one sense they are stories that have appealed to me because they bear some resemblance to the project I have undertaken concerning my own family and the effort I have made, over the past decade, to collect personal documents bearing on them.[2] Indeed, if there is a common element to the autobiographical texts discussed here, it is not the nature of the families described but rather the emphasis on the process of discovery of the family past as a theme in itself, and on the notion of the recovery of intimate details of family life through systematic research. Inherent in the personal definition of 'family memoir' that I have been slowly forging out of these readings is the concept of a genre that involves the recounting of the search for documentary traces of the author's family members and, sometimes, the times in which they lived as well as the stories of their lives. This corresponds broadly with the terms in which the historian and family memoirist Robert A. Rosenstone describes his own book: 'In its mixture of genres and styles, this work lies somewhere between history, memoir, and autobiography—the multivoiced story of a lineage that includes (as any such work must) the life of the teller of the tale' ('Introduction').[3]

Authors of such narratives usually present the project of researching and writing a family memoir as a highly emotional one, even when the individuals whose lives are being reconstructed are relatives whom the memoir-writer never knew. The resulting books are often projects that required many years of effort (16 years for Bliss Broyard's *One Drop: My Father's Hidden Life—A Story of Race and Family Secrets*, more than 20 in the case of Alexander Stille's *The Force of Things*), thousands of miles of travel, and considerable investments of money. Another element in these stories, in addition to the evocation of the lives of family members and the recounting of the author's search for information is that the authors learn in the course of their searches not only new facts about their family past but also new ways of thinking about themselves. The story in these family memoirs is as much about the author him- or herself as it is about the author's family or about collective history.

What precipitates the writing of a family memoir of this kind? In many cases, it is the death of one or both of the authors' parents. This is most obviously the case with Lydia Flem's *Comment j'ai vidé la maison de mes parents* [*The Final Reminder: How I Emptied My Parents' House*] which is, on a literal level, the story of how she disposed of her parents' possessions and what documents concerning their lives she found in the process. Bliss Broyard's *One Drop* takes its point of departure from the author's father's revelation, shortly before his death, of his partial black ancestry, which he had carefully concealed from his children. The opening line of Nancy K. Miller's *What They Saved* is, 'When my father died, I became a middle-aged Jewish orphan' (3). The death of a parent is, of course, a classic example of a traumatic experience apt to provoke personal reflection. Whether one considers oneself an 'orphan', as Miller puts it, or whether one suddenly realises that, with the disappearance of one's parents, one has become a member of the senior generation of one's family, such a death means a sudden shift in one's position in the family network, and indeed a reconfiguration of the family as a whole.

As these memoirs show, however, the death of parents is also significant in another important way. The bereaved child now has a new relationship to his or her family narrative. The deceased parent is no longer there, either to dictate how that story should be told or to act as a source of information. In that sense, much information about the author's family past is suddenly lost forever. On the other hand, however, a parent's death often gives heirs access to information the parents withheld during their lifetimes, either deliberately, as in the case of Bliss Broyard's father, Anatole Broyard, who had built his life around the project of passing as a white man, or unintentionally, as in the case of the old letters and documents that many family-memoir authors exploit to build their narratives. These documents open new perspectives on their parents, and sometimes on more distant ancestors, provoking questions that can no longer be answered simply by asking parents. The documents often set in motion lengthy searches for more information, obtained by talking to surviving family members, interviewing nonrelatives, and tracking down other documents, such as official records concerning the family's life. The author may limit this search only to parents or extend it to previous generations: Broyard's story reaches back some two and a half centuries to her distant French ancestors who settled in the Caribbean, and in his *The Faithful Scribe*, the Pakistani American author Shahan Mufti links his family to Umar, the second caliph of Islam. Some authors, such as Nancy K. Miller, are only able to come up with fragmentary and frustratingly incomplete information, but others, such as Alexander Stille, inherit or compile substantial archives. In many cases, the author ends up knowing, and sharing with readers, more information about the family than the parents ever could have known.

Although the loss of a parent is the most common starting point for a family memoir, it is not the only one. Daniel Mendelsohn's quest is not a search for information about his own parents, who in fact do not play much of a role in *The Lost*, but for details about the family of one of his great-uncles; Mendelsohn's decision, more than half a century later, to uncover the circumstances of their deaths provides the traumatic starting point for his narrative. In the case of Modris Eksteins, the traumatic loss that drives his search is not that of a person but of a homeland, the Baltic nation of Latvia, where he was born just before his parents began the series of migrations that eventually took them to a new home in Canada. In these cases there is no sudden inheritance of documents to stimulate curiosity, although both authors, who undertook their projects a half century after the traumatic events they describe, were acutely conscious that the last living witnesses to those events were rapidly disappearing as the authors were trying to put together their family stories.

Family memoir of the kind discussed here does not require that the relatives whose lives are being reconstructed be exceptional individuals. For the most part these authors depict their relatives as ordinary people with their individual quirks and shortcomings. What justifies these stories is the light they shed on family relationships and, often, on the ways in which history shapes ordinary people's lives, rather than the specific qualities of the subjects themselves. As Alexander Stille puts it,

> Whether we know it or not, we are born into history and pass through a stretch of it. The zeitgeist rubs off on us and leaves its telling signs like the carpet fibres at a crime scene. [...] Our lives have meaning—above and beyond our individual qualities—because we are part of and express the times in which we live. (34)

At the same time, however, these memoir projects are justified by their insistence on the importance of the stories of individual lives. The purpose of Daniel Mendelsohn's *The Lost* is not simply to tell the story of the Holocaust one more time but to make it clear that the six otherwise unknown relatives of his family 'were, once, themselves, *specific*, the subjects of their own lives and deaths, and not simply puppets to be manipulated for the purposes of a good story' (502).

African American author Alex Haley's *Roots*, published in 1976, established a pattern for such memoirs, but Daniel Mendelsohn's gripping *The Lost: A Search for Six of Six Million*, in which the author, a well-known American classicist and essayist, describes his obsessive search for information about members of his mother's family who perished in the Holocaust, has reinvigorated the genre and given it an intellectual respectability that had been put into question by controversies about the accuracy of Haley's story.[4] The power of *The Lost* comes not from any exceptional qualities of its six subjects, about whom Mendelsohn is able to find out relatively little, but from the narrative of Mendelsohn's own search for information, which leads him to travel all over the world, tracking down distant relatives and other survivors from their small town, and from what we learn about Mendelsohn himself in the process of 'living through' this effort with him. Mendelsohn's curiosity about his vanished relatives goes back to his childhood, when older relatives would sometimes become tearful as they remarked, 'He looks so much like Shmiel', the great-uncle killed during the Holocaust. Such experiences gave him a taste of the power of 'postmemory', which Marianne Hirsch has defined as growing up 'with overwhelming inherited memories [...] dominated by narratives that preceded one's birth or one's consciousness [...] to be shaped, however indirectly, by traumatic fragments of events that still defy narrative reconstruction and exceed comprehension' (5).

If *The Lost* has become a model for family memoir, it is because of the passion with which Mendelsohn pursues the traces of these long-dead relatives, the skill with which he narrates the story of his quest, the way in which he draws on his academic training as a classicist in dealing with the larger questions his story raises, and the way in which he unobtrusively demonstrates how the quest for information about his family changes his understanding of himself. At the end of *The Lost*, Mendelsohn reports that it comes as something of a revelation to him when one of his friends helps him understand the relationship between his professional career and his search for information about his family's past. Daniel, it's so obvious. [...] You're a *classicist*, you're a *family historian*. You've spent your whole *life* looking back', his friend tells him (489). *The Lost* is not, however, presented as a holistic portrait of its author; it tells readers nothing, for example, about Mendelsohn's coming to terms with his homosexuality, the theme of a subsequent autobiographical article ('American Boy').

Mendelsohn's memoir has become a classic of family-quest literature, cited by other authors who have undertaken similar projects, such as Nancy K. Miller.[5] It is nevertheless just one of many inquiries into family history conducted in recent years by authors of Jewish ancestry whose families were directly affected by the events of the Holocaust. Other examples include the Belgian psychologist Lydia Flem's *Lettres d'amour en héritage*; Mark Raphael Baker's *The Fiftieth Gate*; most recently, Alexander Stille's *The Force of Things*; and the most famous instance of the genre, Art Spiegelman's *Maus*, which is at once the author's reconstruction of his father's wartime experiences, the story of his relationship with his father, and the story of his own effort to convert those stories into a narrative in comic-book format. There are also, however, a number of family memoirs that concern Jewish families whose lives had no connection with the Holocaust, such as Nancy K. Miller's *What They Saved* and Robert A. Rosenstone's *The Man Who Swam into History*. The Jewish preoccupation with ancestors, embodied in the religious ritual of Yizkor, the thrice-yearly ceremony in which Jews specifically commemorate deceased family members by name, is not limited to Holocaust survivors and their descendants, and the command 'to remember' is central to the practices that have kept the Jewish people in existence for over three millennia.[6]

The concern for family memory is not unique to Jewish authors, of course. Some family memoirs concern the trauma of World War II, but from perspectives very different from those of Holocaust victims, such as Modris Eksteins's *Walking since Daybreak: A Story of Eastern Europe, World War II, and the Heart of our Century*, and others are by authors whose family pasts are entirely different, such as Bliss Broyard's exploration of her father's family's African American heritage or Shahan Mufti's story of his family's interaction with the history of Pakistan. One could even include in this category Barack Obama's *Dreams from My Father*, which ends with the young Obama's 'roots' trip to Kenya to meet his deceased father's family, although the search for family information is not the book's main theme. Indeed, one of the attractions of the family-research memoir is its demonstration that the project of understanding oneself through understanding one's family past is one that has universal meaning, offering a way of transcending differences between groups.

Whatever ethnic backgrounds they come from, the authors of the memoirs in this group, whether they are professors or not, share a commitment to a painstaking kind of scholarly research, with close attention to detail of the kind commonly associated with academic monographs. Like conscientious historians, these authors guide their readers through the process of collecting and contextualising the documents they use, of reconciling discrepancies, and of recognising the blind spots and unanswerable questions left by them. Alexander Stille writes,

> I scanned documents and photographs, transcribed letters, dated, organized, and inventoried them, photographed my father's articles from Italian libraries,

> created a personal archive in which the rich and messy complexity of their [my parents'] lives was reduced to a few gigabytes of digital data and kept neatly inside my hard drive. (429)

Lydia Flem says, 'I transformed my garage into an archival depository, I put up metres and metres of shelving, arranged the boxes, binders, files that they themselves had saved over the years' (*Comment j'ai vidé* 51). Most of these books include family photographs, which are, of course, a common feature in memoirs and autobiography in general. These memoirs, however, often also feature reproductions of non-pictorial family-history documents, such as the sketch of a family tree that also appears on the dust jacket of Miller's book. Alison Bechdel's *Fun Home* includes painstakingly hand-drawn reproductions of letters from her father and entries from her own diary.

Whereas academic historians usually limit themselves to analysing the content of the archival sources they use, family memoirists often describe them physically, sharing with readers details such as the colour of the paper and ink on old documents. Like archaeologists, memoirists often describe the circumstances in which particular items were found: hidden in a drawer, stored in a box tied with a coloured ribbon, buried in a midden of hoarded personal detritus, unexpectedly stumbled upon in a register in a public archival repository. When memoir authors have obtained information by conducting interviews, they often go far beyond the protocols of oral history in evoking these encounters. Alexander Stille recorded his mother in a hospital room, where she was undergoing treatment for brain tumours. The cortisone medication she was receiving 'worked something like a truth serum', Stille writes, unexpectedly facilitating his research (388). The lovingly re-created conversations Daniel Mendelsohn had with elderly Jewish survivors from Bolechow include details of the clothes they wore, the decor of their apartments, and the food they shared with him. The sequence in which they made discoveries about their family is often a key element in these narratives as well. It is rare for memoir authors to uncover the evidence about their family past or their parents' lives in an orderly fashion; more often, as memoir authors' quests unfold, they are forced to shift forward and backward in time, creating complex narratives in which the course of the subjects' lives and the course of the author's reconstruction of those lives intersect in unpredictable ways, with the progress of the latter often changing the understanding of the former.

In academic historical scholarship, the kind of meticulous attention to the process of research that characterises these memoirs is often categorised as 'positivism', a determination to demonstrate that the author has done everything possible to let the facts speak for themselves, in an objective fashion. Positivist scholarship, it is often said, minimises the role of the investigator; in the natural sciences and some of the 'hard' social sciences, scholars are taught to report their findings in the passive voice, using phrases such as 'it was observed', as if observation was a completely impersonal process. In family memoirs, however, the obsession with process has almost the opposite function.

The description of the author's research process keeps the memoir writer and his or her emotional subjectivity in the centre of the story. Daniel Mendelsohn's peregrinations across four continents to interview survivors; Bliss Broyard's trips to archives and her round of family reunions with previously unknown relatives in New Orleans, Los Angeles, and elsewhere; and Alexander Stille's epic deconstruction of the mountains of hoarded trash in his aunt Lally's New York apartment share equal billing in their memoirs with the facts they uncover about their relatives as a result. Lydia Flem suggests that family memoirists' obsession with documentation is not just a matter of reassuring readers that the stories recounted in their books are true, but a way of stabilising themselves, of limiting the fantasies about their families' lives that might otherwise haunt them. 'Pin down the dates, note the facts, regard their truth as a reality, not only as a terrifying phantasm, devoid of sense', she writes (*Comment j'ai vidé* 74).

Family memoir authors' concern to draw readers into the details of their searches for evidence and to bring them as close to their sources as possible coexists with an acute and frequently emphasised consciousness of the difficulties in extracting anything like truth from such documents. Robert Rosenstone in fact argues for a postmodernist scepticism about the possibility of objective knowledge. 'The reality of the past—national, familial, personal—does not lie in an assemblage of data but in a field of stories—a place where fact, truth, fiction, invention, forgetting, and myth are so entangled that they cannot be separated', he writes ('Introduction'). Pouring through an ancestor's painstakingly compiled family genealogy, a document purporting to connect the family with one of the founders of Islam, Shahan Mufti realises that 'his story may have been partly imagined, partly true' (Ch. 9). Confronted with two conflicting stories about the fate of one of his relatives who perished during the Holocaust, Mendelsohn writes,

> On the one hand, it was disconcerting; I was beginning to become aware of how fragile each story I heard really was. [...] On the other, I felt the strange exhilaration you can feel when faced with a particularly challenging mystery story. [...] What had happened to Uncle Shmiel, then? (*Lost* 303)

Even as they mimic the procedures of the most scrupulous scholars, these family memoirists simultaneously recognise that that the project of writing about one's own family can never be only a matter of constructing an objective historical narrative. Research into a familial past is an intensely emotional process, a point brought home particularly in illustrated memoirs such as Spiegelman's *Maus* and Bechtel's *Fun Home*, in which the facial expressions and body language in the authors' pictures of themselves at moments when they were confronting difficult aspects of their parents' lives speak louder than words. Bliss Broyard provides a striking verbal demonstration of the emotional element in family-history research in her description of the day on which she realised that documents in an archive outside of New Orleans proved that one of her mixed-race great-grandmothers had been, not a slave, but a slave-owner:

> I raced back to the city, across the twenty-four-mile bridge, riding all the way in silence, shaking my head every so often and swearing under my breath. The leaps of logic I'd made and the clues I'd missed in researching Marie's genealogy were suddenly distressingly clear. The thought of how blinded I'd been in my obsession to find a slave ancestor made me feel sick with shame [...]. In a few short hours, I'd gone from believing that my great-grandmother was born a slave to discovering that she'd grown up in a family of black slave *owners*. (267, 268)

The product of all this effort is both a biography or a history, and also a kind of autobiography. In retracing the story of her father's family, for example, Broyard provides a history of the community of 'coloured people' in New Orleans going back to the beginning of the nineteenth century and a depiction of the choices facing light-skinned people of mixed ancestry like her father in the mid-twentieth century, at the time when he consciously decided to separate himself from that heritage. As Broyard emphasises, this is a story that is of broad concern in understanding race relations in America, but she was only motivated to learn about it once she decided to try to achieve a better understanding of her father. And even though her book is ostensibly about her father's 'hidden life', *One Drop* is just as much the story of its author as it is the story of her elusive parent. The contrast between her eager acceptance of the fact that she was part black and her father's elaborate effort to deny his ancestry, her laborious efforts over 16 years to locate relatives she had never met during her father's lifetime and to reconstruct the family's more distant past, and her frustrated realisation that the branches of the family that her father had separated himself from were not really ready to welcome her into the clan, are all parts of her own life narrative, events that took place after her father's death.

This element of self-narration is perhaps most obvious in the illustrated memoirs of Bechdel and Spiegelman, in which the authors appear as characters in the story, and in Rosenstone's, where the author chooses to portray himself under the pseudonym 'Rabin'. Bechdel's varied depiction of herself, sometimes as a small child, sometimes as the young adult she was when her father committed suicide, highlights the importance of her own story within the family story she tells. Alexander Stille, who chose to go into his father's profession of journalism and to specialise in writing about Italy, the country where his father had grown up and whose language his father was most comfortable with, could hardly avoid discussing the relationship of his own career to his family background.

> Any dime-store psychologist could see that this was a form of identification with my father: coming to love Italy had given me ways of loving my father and allowed me to pick a profession that was almost synonymous with my father. (416)

The self-portraits constructed in these narratives of search for family are complex and sometimes contradictory. In one sense, these authors are clearly seeking for connection with the families from which they come; in another sense, the very writing of such a memoir emphasises a sense of distance and separation, for if the author and his or her ancestors—usually the parents—had truly been close, the immense labour that family-memoir authors go through to understand their family heritage might not have been necessary. These memoir authors do not wish to portray themselves as eccentric or extravagant personalities, but some readers might find their devotion of years of effort to reconstructing the minor details of their relatives' lives excessive, the sign of a personal hobby allowed to get out of hand.

The sharing of the intimate details of family members' private lives with a public of anonymous readers obviously has its problematic aspects, which have been explored by a number of writers, including Nancy K. Miller, herself the author of a family quest memoir. 'It is not my wish to do harm', she wrote in an essay published some years before *What They Saved*, 'but I am forced to acknowledge that I may well cause pain—or embarrassment to others—if I also believe, as I do, in my right to tell my story' ('Ethics' 157). John D. Barbour, another contributor to the volume on *The Ethics of Life Writing* in which Miller's essay appeared, noted that family memoirs tend to move from negative initial judgments of parents to 'a more nuanced assessment, and to tentative, partial forgiveness'. Further, this 'socially approved goal of forgiveness may authorize a writer's exposure of family secrets and emotional outpouring, which would otherwise look like vindictiveness or whining' (95). Lydia Flem is highly conscious of the fact that her project started with an act of 'indiscretion'. She describes this as the 'obligation to disregard all the rules of discretion: to go through personal papers, open handbags, read mail that wasn't addressed to me'. Her first book, which focused on her own thoughts and feelings after her parents' death, was arguably the story of a part of her own life, but her second, based on the love letters her parents exchanged before their marriage, obviously raised different questions. 'Wasn't I going to read things that should have remained unknown and secret for me', she asked herself? (*Comment j'ai vidé* 27). When she described her project to friends, she became aware of their 'strained silence' (*Lettres* 29–30). Flem managed to convince herself that her project was justified by the fact that she had been 'born from this correspondence'(29–30), but she had no way of knowing how her parents might have felt about the use she made of it. Having devoted two volumes to her engagement with her parents' lives, she went on to publish a book about an even more sensitive topic: *How I Separated Myself from My Daughter and My Sort-of Son* [*Comment je me suis séparée de ma fille et de mon quasi-fils*].

Alison Bechdel's posthumous 'outing' of her father, including the story of how he was once arrested and narrowly escaped prosecution for pursuing teenaged boys, may be justified as an effort to change the cultural attitudes that, in her view, stunted his life and led him to withhold the truth about himself from his own children. Bliss Broyard's exploration of her father's struggle to conceal the truth about his ancestry obviously speaks to the continuing destructive power of racial prejudice in American life. It is less clear what larger purpose is served, however, by Alexander Stille's detailed reconstruction of his mother's lively social and sexual life as a young woman or the story of the sunburn his father got from spending too much time having sex with her on a Caribbean beach. Stille's justification for writing about his parents is typical of family memoirists whose subjects are no longer living: he knows he depicted them 'differently than they would have portrayed themselves. But my parents were dead and their story was also mine' (431). He was more troubled by the reactions of living family members who appear as characters in the story. When he showed his manuscript to his father's sister, his aunt Lally, she responded by sending him her own version of her life story, 'a counter-narrative, written from her own point of view', written in the third person, and demanded that he include it in his text. Stille refused and in fact compounded the sin he had committed in his aunt's eyes by describing his debate with her. She was not the only relative to object to his project; he came to realise

> that the very act of my writing about our family, which was of course also *their* family, was deeply destabilizing in and of itself. [...] All of us, after all, develop a conception of the world and our place in it, work out a narrative of our lives, and these are things we need vitally to live. [...] Having someone else come along with his own narrative—especially if it is published and becomes a kind of official version—is deeply destabilizing. (436, 435)

The relatives of the authors of such memoirs may well be, as Stille suggests, more resentful of than grateful for them. But what is the appeal of such narratives to readers who have no connection with the authors or their families? In many of these narratives, the authors begin, as John Barbour remarks, with a critical moral view of their parents, who are blamed for concealing essential aspects of their identities from their children, as in the case of Broyard and Bechdel, or for their irresponsible and sometimes abusive behaviour, as in Stille's story. As they learn more about their parents' lives, however, most of these authors arrive at an 'ethical assessment that takes account of both what was within the parent's voluntary choice and the ways in which he or she was constrained, pressured, or determined by things outside his or her control' (Barbour 91). Such stories may have meaning for readers whose own parents were less than perfect, a category that probably includes much of the human race. Not all the memoirs discussed here fall into this pattern, however. Parental shortcomings are not a theme in Modris Eksteins's or Shahan Mufti's accounts of family histories shaped by the fates of the countries of their ancestry, nor are they an issue in the warm and admiring portrayal of her parents that Lydia Flem provides.

The specific issue that ties this group of family memoirs together is not simply the effort to reach an equitable moral perspective on one's ancestors, but the role of the quest for meaningful information about those ancestors in the author's life. Although there are several non-academics among the authors I have discussed here—Bechdel and Spiegelman are professional comic artists, and Flem is a psychologist—the majority of them are either academics or journalists who have systematically applied their professional skills to the search for family information. As she crosses the country in hope of finding documents about an uncle she never knew, Nancy K. Miller describes herself as pursuing 'an academic's fantasy, chasing the grail of the killer archival find' (*What They Saved* 132). The narratives make it clear how much effort the authors invested in doing their research, even if most of them prefer to emphasise the role of luck and coincidence in leading them to critical finds rather than crediting their success to dogged persistence. For many of these authors, family-history projects were clearly diversions from the more 'important' work to which most of their careers have been devoted, which no doubt accounts for why some of them took a decade or more to come to fruition; Eksteins mentions, for example, that he wrote *Walking since Daybreak* in place of a historical monograph about Europe in 1945 to which he had originally committed himself. But these narratives are vindications of the values of scholarship, even when applied on the micro level of family-history research. Knowing that memory alone is unreliable and that emotion can cloud the understanding of documents, these authors deliberately construct stories that emphasise the evidential basis for their assertions about their families. By implication, they suggest the importance of situating even the most intimate and subjective details of personal experience in a framework subject to procedures of documentation and verification. In an age all too prone to assume that memory, with its emotional appeal and its link to fiction, is more important to human life than the factual record, projects like these reassert the value of a more critical view, even of the personal past.

Just as writing family memoirs has been a diversion for many of the authors I have discussed here, reading these research-driven family memoirs has diverted me from my own project, which has also been on hold because of questions about the propriety of using some of the family papers I have accumulated and because of the time and energy I have invested in other scholarly projects. It is also intimidating to think of entering the same literary arena as writers like Daniel Mendelsohn and Alexander Stille, and I have no talent for drawing to permit me to even dream of accomplishing what Art Spiegelman or Alison Bechdel have done. In exploring these projects by other authors, however, I have also been able to articulate more effectively the sense of my own project and the meaning it could potentially have for others beside myself. I have no doubt that there is a story, indeed several stories, in the files of family life-writing documents currently sitting quietly in my study. I now recognise, however, that the authors of family memoirs must be willing to open themselves to the same kind of scrutiny that they impose on the family members about whom they write. It is a daunting prospect.

Notes

[1] This discussion is based on the following family memoirs: Mendelsohn, *Lost*; Miller, *What They Saved*; Flem, *Comment j'ai vidé* and *Lettres d'amour en heritage*; Baker; Stille; Eksteins; Broyard; Spiegelman, *Maus*; Bechdel, *Fun Home* and *Are You My Mother?*; Rosenstone; Mufti. The first volume of Flem's family memoir has been translated into English as *The Final Reminder: How I Emptied My Parents' House*

(London: Souvenir Press, 2005), but citations in this paper are my translations from the French edition. Little noticed in the English-speaking world, Flem's book has been translated into nine other languages, including Catalan and Basque.
[2] See Popkin.
[3] References to the works by Mufti, Rosenstone, and Stille are to e-book versions; page numbers may not correspond to those in the print editions of these works.
[4] On the importance of *Roots* and the controversies it caused, see Weil (192–8).
[5] Miller, *What They Saved*; see also the reference to one of Mendelsohn's essays on the subject of memoir in Bechdel, *Are You My Mother?* (10).
[6] See Yerushalmi's classic essay.

References

Baker, Mark Raphael. *The Fiftieth Gate: A Journey through Memory*. Sydney: HarperCollins, 1997.
Barbour, John D. 'Judging and Not Judging Parents'. *The Ethics of Life Writing*. Ed. Paul John Eakin. Ithaca, NY: Cornell UP, 2004. 73–98.
Bechdel, Alison. *Are You My Mother? A Comic Drama*. Boston: Houghton, 2012.
Bechdel, Alison. *Fun Home: A Family Tragicomic*. Boston: Houghton, 2006.
Broyard, Bliss. *One Drop: My Father's Hidden Life—A Story of Race and Family Secrets*. New York: Back Bay, 2007.
Eksteins, Modris. *Walking since Daybreak: A Story of Eastern Europe, World War II, and the Heart of Our Century*. Boston: Houghton, 1999.
Flem, Lydia. *Comment j'ai vidé la maison de mes parents*. Paris: Seuil, 2004.
Flem, Lydia. *Comment je me suis séparée de ma fille et de mon quasi-fils*. Paris: Seuil, 2009.
Flem, Lydia. *Lettres d'amour en heritage*. Paris: Seuil, 2006.
Hirsch, Marianne. *The Generation of Postmemory: Writing and Visual Culture After the Holocaust*. New York: Columbia UP, 2012.
Mendelsohn, Daniel. 'The American Boy'. *New Yorker*, 7 Jan. 2013. <http://www.newyorker.com/magazine/2013/01/07/the-american-boy>
Mendelsohn, Daniel. *The Lost: A Search for Six of Six Million*. New York: Harper Perennial, 2006.
Miller, Nancy K. 'The Ethics of Betrayal: Diary of a Memoirist'. *The Ethics of Life Writing*. Ed. Paul John Eakin. Ithaca, NY: Cornell UP, 2004. 147–60.
Miller, Nancy K. *What They Saved: Pieces of a Jewish Past*. Lincoln: U of Nebraska P, 2011.
Mufti, Shahan. *The Faithful Scribe: A Story of Islam, Pakistan, Family, and War*. New York: Other, 2013.
Obama, Barack. *Dreams from My Father: A Story of Race and Inheritance*. New York: Three Rivers, 1995.
Popkin, Jeremy D. 'Autobiography in the Family'. *a/b: Auto/Biography Studies* 25 (2010): 172–85.
Rosenstone, Robert A. *The Man Who Swam into History: The (Mostly) True Story of My Jewish Family*. Austin: U of Texas P, 2005.
Smith, Sidonie, and Julia Watson. *Reading Autobiography: A Guide for Interpreting Life Narratives*, 2nd ed. Minneapolis: U of Minnesota P, 2010.
Spiegelman, Art. *Maus*. New York: Pantheon, 1996.
Stille, Alexander. *The Force of Things*. New York: Farrar, 2013.
Weil, François. *Family Trees: A History of Genealogy in America*, Cambridge, MA: Harvard UP, 2013.
Yerushalmi, Yosif H. *Zakhor: Jewish History and Jewish Memory*. Seattle: U of Washington P, 1982.

Voices in Movement: Feminist Family Stories in Oral History and Sound Art

Margaretta Jolly

Voices in Movement, a sound installation directed by artist Lizzie Thynne with music by Ed Hughes, draws on memories recorded for Sisterhood and After: The Women's Liberation Oral History Project. Both the installation and oral history foreground family stories as central to feminist politics, though in diverse and shifting ways. As the producer of the installation and director of the oral history project, I explore how these representations of the family emerge, as well as the differences between oral history and sound installation as forms of life story-telling.

So this was a family secret. We, as children, we had no idea about this. My mum and dad gave us this whole picture of the family as this wonderful religious unit, you know, everybody doing their duty and nobody straying and nobody having bad morals and all that stuff, you know, and it didn't come out for about, well, until I think it was the twenty-fifth wedding anniversary. (Voice of Barbara Jones, excerpt from Thynne and Hughes)

Yes, I mean I was a bad girl. I have an overwhelming sense of being a bad girl. [...] My sister's sense was that I was made the bad girl. [...] I was also the only important one [...] that she was the also-ran. We are very, very close. But it's been, you know, it's had its ups and downs, and her and my sense of our family life tallies completely, which has been very important to me, and, and which I knew nothing about until I started to get ill. (Voice of Barbara Taylor, excerpt from Thynne and Hughes)

The sounds of two women's voices intertwine, full with age, strong with conviction. But the stories they narrate are of childhood confusion, of traumatic secrets and a puzzle over love. One voice is inflected with the Yorkshire accent of the English industrial North, the other Anglo-tinged West Coast Canadian, and they are in the middle of the tale when they float out of the sides of the dark gallery. After a few minutes a pale family, coiffed 1940s style, appears on a screen at the front. They are picnicking and the sun illuminates the Chevrolet, bounces off the tartan rug. A little girl looks up from playing with her cup and squints at us before the image dissolves. The voices pulse on.

> Because, her sense of me was that there was the three of us, my, my parents, especially my mother and me, and that she was the kind of outsider, [...] and that we were kind of, bound together in our way of being a family and understanding a family, and to which she felt no affinity. So, when everything started to kind of *crack* for me, she was able to start speaking to me about her sense of the family, which was *stunning* to me. (Voice of Barbara Taylor)

> They had to tell Maggie, who was the child in question, that actually, she was conceived quite a long time before they got married. Honestly, we were so—[laughs]—we just, we were so outraged really, I certainly was, that they'd presented this picture of the idyllic family to us. (Voice of Barbara Jones, excerpt from Thynne and Hughes)

These fleeting images soon disappear. But the voices continue with their family stories: remembering, explaining, questioning.

Family Stories, Feminist History

Family stories are fundamental to life narrative in all media. However, oral history methods have been particularly powerful in capturing them for the collective record. As a method that has been central to social and popular history where documentation is weak, interviews capture changing worlds of childhood and adolescence and the kitchens and bedrooms within which they unfold—the experience of shift work or housework. Classically, they tell of inheritance, care, rivalry, as well as dreams of upward mobility through generations. The eminent English oral historian Paul Thompson argues that the method's inclusion of family stories has allowed major revisions in our understanding of history, through the exposure of dimensions such as, for example, the economic role of extended families during industrialisation (Thompson 300–4). The British National Life Stories collection begins with detailed questions about parents, childhood, home life, and relatives, in the belief that this creates the most contextualised and flexible archive for future study. Their recent 'Oral History of British Science', for example, 'revealed insights into the role of spouses, sacrifices made in family life, the "long hours culture" of a career in science, and how leisure activities are often an extension of scientific work' (Perks). Indeed, most handbooks on oral history interviewing concur that family stories are at the heart of much oral history interviewing across the board (Yow *Recording* 253–79).

The oral historical family record is dramatically enriched by a feminist perspective. Interviews recording other hitherto hidden histories of women's domestic labour and sexual control provide profound insights into gender relations in terms of women's economic dependence, social isolation, and physical or sexual abuse. This is why feminists have argued so strongly against its romanticising in ideologies of marriage and home. And yet, as the debate following Mary McIntosh and Michele Barrett's 1982 *The Anti-Social Family* proved, family is also the subject of intense theoretical and political disagreement within feminist circles (Barrett and McIntosh 'Ethnocentrism'). Is the family to be reformed, dispersed, collectivised, abolished? The debates opened up questions about the myriad definitions and forms of family life. The family against which the white middle-class core of the Western movement fought in the 1970s was the nuclear version, shrunk through industrialisation, polarised by capitalist patriarchy at a particular reactionary mid-twentieth-century moment. Furthermore, life as a happy housewife remains for many an unattainable ideal, even as for others it is a false promise. Black feminists have also proposed that family can be a resource against other oppressions of race and religion, where they have been undermined, divided, or impoverished through war, migration, inequality, and prejudice (Amos and Parmar). Family is the place of transmission for identities that feel marginal, on the defensive even as it is also the mechanism for reproducing pinkified girls and bluish boys and national pride of the most virulent sort. In recent years, in some cultures, families are being redefined through the queering of parenthood and even grandparenthood as those of minority sexualities have won the right to have children (Vaccaro).

The transformation of family stories within and alongside feminism is reflected in their changing forms. Marianne Hirsch has argued that familial structures are basic to narrative and, conversely, that narrative structures underlie traditional conceptions of family. In *The Mother/Daughter Plot: Narrative, Psychoanalysis, Feminism,* Hirsch designates the bearing and regeneration of life as a formal concern that has found new expression with the cultural enfranchisement of women (3). Iterative, circular, and open-ended structures mimic the more relational life story of maternal rather than paternal identification. Such 'spiral' family plots certainly seem present in feminist memoir writing, as Helen Buss, Nancy K. Miller, and Tess Cosslett, Celia Lury and Penny Summerfield, amongst others, have shown. Today, the proliferation of different versions of the parent-child plot—Alison Bechdel's visual father-quest, for example—as well as misery memoirs, 'Who do you think you are' documentaries, digital life story curations, and the genealogical craze capture and promote fresh, postpatriarchal concepts of family, even as fascinations with identity, inheritance, the private, the domestic, and the sexual endure (Bechdel; Couser; Lynch).

In planning an oral history of the women's liberation movement (WLM) in the United Kingdom, we—a team that included a curator, a filmmaker, academic cultural historians, as well as activist advisers—were well aware of our need to approach family as a political question. The resulting *Sisterhood and After* project, which took place during 2010–2014—captured 60 life history interviews with core activists from the 1960s–1980s across the United Kingdom for archiving at the British Library. Following the library's method, we averaged seven hours' interviewing time per person and asked questions across the life span in chronological order. As feminists, we began with a question about the origins of our interviewee's name, which revealed patriarchal genealogies even in the moment of introduction. We continued by asking about the interviewee's forbears beginning with her mother, and invited further information on extended family, siblings, lovers, and friends. We learned to hear traumatic family stories with neither judgment nor shocked silence. We also discovered the importance of feminist family-making in activists' mid-late lives. We ended the interview by asking the interviewee to compare her life to her mother's. This aimed to measure the change in women's opportunities across generations, but invariably emphasised the deep sense in which an individual story is a family story, whatever its emotional flavour or political definition (Jolly, Russell, and Cohen).

The resulting oral histories amply demonstrate the complexity of feminism's relationship to family. The capaciousness of the method enabled the interviews to reveal that, even at the time of communal living experiments, strong connection with families of origin endured. The distinction between patriarchy as a system and the varied reality of real fathers was obvious.[1] Many mothers were apparently more conventional than the fathers—a pattern not uncommon in feminist movements, where often it is the educated father who encourages the daughter's politics. Sometimes, mothers became feminists themselves but not always in a way that solved tense relationships with their activist daughters.[2] At the same time, the interviews also challenge any simple view that family is a haven against state, race, religious, or class oppression.[3] Family is confirmed as a formative site of struggle that is specific to women across diverse locations—and yet quite clearly, family, including the nuclear family often survives as a deep emotional and material resource. In trying, therefore, to understand the part played by family relationships in the making of a feminist, as Olive Banks put it of the suffrage movement, '[I]t is obvious that no simple explanation will suffice' (33). The next section of this paper focuses on two of the interviews and the stories they revealed. They are those of Barbara Taylor and Barbara Jones.

Hearing, not Reading

Though life writing by feminists has elegantly explored many of these themes, revising and rewriting in a way that oral history cannot, it is not able to capture the full power of the voice in the telling. We can see this simply by contrasting Barbara Taylor's recent memoir with her interview with me for *Sisterhood and After*. *The Last Asylum* tells the painful story of Taylor's breakdown and treatment as a voluntary inpatient in a London hospital in the late 1980s, set in a larger argument about the history of treatment for mental health. It also quietly sets out the irony that her brilliant career and socialist family of origin did not prevent years of misery and self-doubt, in which her almost miraculous recovery was, she relates, due to her feminist friends, sister, and psychoanalyst. At one point in the interview, Taylor reflects on writing her memoir as part of a general coming out about this 'shameful' past:

> [T]hroughout [...] this interview, I've been conscious of forms of censorship, inevitably, and because I'm someone who has had long experience in talking about myself, I mean 21 and a half years [...] of psychoanalysis, which was uncensored, but has given me a capacity for self-censorship, I hear what I'm saying before I say it a lot of the time, but I truly don't know what I'm going to put on the page, I've got reason to hesitate. [...] I think, you know, I've set out to tell this story, and if I don't tell the bits that are most uncomfortable for me, then, people won't understand. [...] I don't know why I need people to understand, I truly do not, except that I think that there is, there is a [...] shame that surrounds mental illness...[4]

Few are as perfect a stylist as Taylor. Her memoir was named BBC Radio 4's Book of the Week, an honour that added to the acclaim she had already received as an intellectual historian.[5] On receiving the oral history transcript, then, it was unsurprising that she declared herself horrified by her 'habit of repetition', though clearly she positions the oral history interview as several degrees more 'censored' than a therapeutic conversation. But anyone who listens to the recording itself will be mesmerised by her self-reflective drawl, indeed by precisely that repetition, for it embodies unfolding thought, aurally and sometimes with confident insistence. The theatrical drawing out of vowels, the sudden rushes, the drop at the end of the sentence—these convey practice in public speaking, where repetition is also the first rule of style. By listening to her voice, however, we can also hear that the confident gestures, the generalisations, the 'we' of 'the human condition', are in counterpoint to her vulnerable, intimate past self.

Taylor speaks of an early family life to which she clearly attributed her adult breakdown. She grew up in western Canada, the child of a lawyer mother who became the city's first magistrate judge, was a member of the antinuclear Voice of Women and also the head of the Commission of Human Rights in Saskatchewan. Her family were important figures in the Jewish community in this small city. 'Mum' had gone to university at age 15, but thence disappointed her ambitious parents by falling in love with an older, Gentile communist. This driven mother, who had rejected her own family convention, in turn drove Barbara, for whom no academic or professional achievement could be good enough. The pressure was further stoked by Taylor's 'ferociously bright' Welsh emigrant father who had fought in the Spanish Civil War as a young man. Just as her mother was not, in fact, oppressed by poverty or the demands of housework— Mennonite girls were brought in to care for Taylor and her sister—the oppression was psychological, to do with a toxic mix of sexuality, inherited ambition, and contrary political messages as her revolutionary hero father bullied, philandered, and needed. As Taylor summarises, 'Dad was a domestic tyrant of the old school, Mum his adoring slave' (Taylor 16).

Taylor's childhood story was not especially dramatic. Like many activists, she was relatively privileged and joined the women's movement in her twenties when she had time, brains and enthusiasm to spare, and the support of a scholarship to London. The early 1970s passed in a whirl of intellectual protest against the patriarchal status quo, in which a Lacan reading group was a high point. These interests later fed into an ability to theorise but not cure a devastating emotional illness, an irony that is palpable. She vividly describes the socialist version of the 1950s climate of gender/sexual repression that formed her, over-determined by ethnic and class aspiration. This painful demythologising of the family is metonymised in her memoir's opening pages, where she describes coming across an old copy of the Museum of Modern Art's 1955 catalogue *The Family of Man*, which she had loved as a child. In her early thirties, she suddenly sees that, amongst the idyllic photographs of children around the world, other children are pictured alone; people are screaming or starving. She realises suddenly that she has idealised her heroic radical parents (Turner 12–13, 16–17). Her subsequent fragmentation, caused in part by the psychoanalysis she sought to treat it, is mimetically rendered through self-exposing remembered dialogues from her therapy. These show her regression, as she furiously searches for the nurture from her doctors that she felt she did not get from her parents, yet also her return to life and, in that process, to forgiveness.

As I listen to her interview, a similarly powerful story of past pain unfolds, despite my peppering prompts for names and dates for the archive. Across my questions, we hear her constructing a narrative that contains as well as exposes the family story.

> Yeah. No, I mean, I mean my breakdown and, and the, the therapy and so on, in the end had a very very good impact on my relationship with my parents. You know, the … Who are our parents to us, you know? We carry inside us a version of our parents, or rather, the people who happen to be our parents, you know, the parent figure, is a kind of, it's a version isn't it, of the actual human being. And, in the case of my parents, I mean they were very distorted figures inside me I think, and, made quite frightening and monstrous by my own difficulties. They weren't great at parenting, but they weren't anything like as bad as the version I think that I carried around inside me.[6]

This spoken story is actually less self-exposing than the memoir's portrait of her drunken, bullying, babyish former self, which was perhaps easier after all to release in the aloneness of writing. And yet hearing Taylor is moving, even seductive, in a different way, as an authoritative woman's voice delicately counterpoised against the upsetting history, embodying the survival she explains. Moreover, Taylor speaks here of her love for her sister much more clearly than in the book—as she does of the individual friends and now her current partner and partner's children. This brings out the possibility of remaking the family in adulthood, and choosing its configuration, here with intensely happy results.

> [I]t was absolutely revelatory. And then became, as I say, very important for me in helping me to kind of reinterpret family life, in a way that actually made sense to me instead of the kind of, the myth in a sense that I was trying to sustain inside my, inside myself.[7]

Taylor's spiralling relationship to family, to my mind, reveals new elements of feminist life that complement the memoir and underline the way family is refigured at different life stages. In answer to my final question, 'How does your life compare to your mother's?' Taylor responded that without doubt hers had been luckier, despite the misery that her mother had put her through.

The second interview, with Barbara Jones, offers a different argument for the value of oral history, in the classic sense of capturing a story from someone who probably would not have written it. Yet here, too, aurality is significant in its own right, for Jones represents the tiny 1% of women who work as builders and roofers in the United Kingdom, a percentage that has not changed since she trained in the mid-1980s.[8] Indeed, her work 'on tools' is very unusual even within the feminist community in contrast to the numerous social workers, educationalists, and policy workers. Her descriptions of training in carpentry, roofing, plumbing, and building are testimonies in their own right.

> I liked the meticulousness of it, I liked the accuracy of it, I liked using my body to be skilful. I liked to be able to make a straight saw cut and to make a joint that fitted perfectly. It gave me pleasure to be able to make complicated joints and dovetail joints and for them to fit perfectly together and to look good. And, you know, when something looks good most people don't even know the work that's gone into it because they're that good, but you know and other carpenters know, and that's what I loved.[9]

Her voice—light, articulate, punctuated with laughter—is captivating in its own right, where a woman's physical presence and ability is in itself so politically significant. And it is of a piece with her narrative, which consistently positions her life and relationships in terms of the choices governed by radical feminist and lesbian philosophies of women's autonomy. 'I think an awful lot of women have been self-employed in construction because that way you've got freedom,' she explains, 'you can run your own life, you can work when you want to, and you can work with who you want to, and that's crucial.'[10] She opens her interview with this point:

> [A] big focus of my life has been to enable women to do things that this culture that we live in stops us from doing and I've had that right from childhood really, from what I learnt as a kid and I still feel very passionate about it now in my fifties. So the things that I've done that have been part of that process, the Women's Movement, being a lesbian, being a feminist, working in manual trades, moving to Todmorden, the centre of the lesbian universe in the UK. I mean all of that and I'm still part of it and I'm still trying to do stuff to make change happen in a positive way for women.[11]

Jones, one of seven children, grew up in Wakefield, Yorkshire, the daughter of parents who worked in local civil service jobs. Money was tight; she remembers seeing her father unhappily leaving for work each morning. Her mother, who had originally been his boss—she was nine years older than he—was now a stay-at-home wife. She had converted to Catholicism for his sake. Her father was passionately religious, and he took the family to live next to the church in the town's Catholic enclave. Jones remembers the white plastic boater and red velvet bow and elastic under her chin as they went to Sunday school. Eventually she got up and out. She studied sociology at Lanchester Polytechnic, became a social worker in London, and discovered a new life and career in carpentry. Her narrative of self-making is suffused with respect for her hard-working parents as well as on-going identification with her siblings' own business enterprise.

Yet it attaches to another plot. Jones speculates that the marriage itself was sparked by her mother's unplanned pregnancy, though this was nigh impossible for her parents to admit. And then, as we sat in the rented interviewing room, she said something more difficult.

> MJ: Well, I want to go back to something you said and just ask if you want to talk about it, which is you said something about what happened to me in the family. If you would like to say more about that, then do.
> BJ: Yeah. Yeah, I do want to talk about it because part of the thing about sexual abuse in families is that women don't talk about it and certainly it was a really, really difficult issue for me to deal with and to voice and it's had a, has had a lot to do with who I am, but also because I was brought up to be honest it just didn't make sense, the secrets that there were in the family.[12]

The words that followed were carefully chosen, without tears or anger so much as an audible wish to testify. She talks about a time when her separatist politics involved preventing male relatives entering her house, and in turn how she was for a period ostracised by her family for her sexuality, and then of her developing sense of having been the survivor of incest. Dealing with this eventually meant confronting her father, and the wreck of her relationship with her then partner. It is one of the saddest moments of the interview.

In her conscious decision to speak out about sexual abuse within her family, and to challenge the hypocrisy even as she celebrated—and evidently continues to celebrate—her family's deeply moral codes of conduct, Barbara Jones places her oral history within a genre that is both a 'coming out' and a survivor story. Hearing such stories is quite different from reading them, even if it is supplemented by transcripts or indeed memoirs. Jones's voice has a magic that, as with Barbara Taylor, is especially powerful in the context of speaking of shame, of secrets, and then of transformation and survival.

From Oral History to Sound Installation

Feminists have long had an affinity for oral history, because of its potential to capture the non-elite experience and because of the intimacy and equality it offers as an interview form. While feminist historians in the 1970s celebrated it as the discovery of voice, or even consciousness raising by the 1990s (Chamberlain 1), Sherna Gluck, Daphne Patai, Ann Oakley, Liz Stanley, and others were foregrounding the interview relationship as a political terrain through which to explore gender identity and difference. But sound is gendered in dimensions that exceed questions of vocal representation. It involves the painful relationship that women have with the visual and the gaze, the lateral and unconscious elements of the aural as repository of the domestic, the sensual and sexual, the bodily place in the world and its connection to others. I have touched on this in the stories of the two Barbaras. Consider, in addition, these moments in other Sisterhood oral histories: Una Kroll, recounting how she was told that women's voices were too high for the radio when she applied to do the BBC's *Thought for the Day* in the late 1970s; Kirsten Hearn's description of finally getting access to an audio version of Germaine Greer's *The Female Eunuch*, more than a decade after it was published for sighted women; Jenni Murray, who has been the voice of BBC *Woman's Hour* for 25 years, recounting her elocution lessons to eliminate her regional and class accent—and the sound of her Chihuahua dogs yapping at my feet throughout the interview; Sheila Kitzinger demonstrating the panting noise a sheep makes when giving birth; Gail Chester laughing as she recounts her appalling sex education; Rebecca Johnson singing the songs of the Greenham Common peace camp; the sound of scholar psychoanalyst Juliet Mitchell's rocking chair; and the thunder outside Mukami McCrum's window in Edinburgh when she talks of her childhood in Kenya. Then there's the intake of breath when you ask almost any of them about their mothers—and indeed, my own, when I was asked in turn.[13]

In responding to these sonic elements of life narrative—both verbal and nonverbal— oral historians may learn much from sound artists for whom audio form is the starting point. Although oral history and sound art originally come from very different circles and places, as historical method on the one hand and musical experimentation on the other, they have grown closer since the 1960s. Reconfigured through the progressive politics of the time, they share interests in everyday life sounds and the history of protest. And a fascination with family stories is as evident in feminist sound art as it is in oral history. While Valerie Yow captured the lives of three generations of women millworkers in North Carolina, Alison Marchant projected the recorded voice of her millworker aunt within the walls of the Barchant cotton mill. While Anne Butler and Gerri Sorenson imagine women's oral history as a 'patchwork' and celebrate their transcripts in quilt form, Suzanne Lacy choreographed ordinary women over 60 to tell their stories to each other while they sat in a grid formation that was, literally, the shape of a quilt (Lacy, Roth, and Mey). Cathy Lane's 1999 *Hidden Lives* juxtaposes field recordings, archival materials, interviews and conversations and synthesised and acoustic instrumental sounds to explore the house as the repository of memories, with women as the curators of those memories. Her more recent work, with Annea Lockwood, *Someone Else Can Clean Up This Mess*, drew upon their own experiences as women working with sound and performance in 1960s and 1980s London, to explore the emergence of a feminist subjectivity. The discussion between the two artists was interwoven with slides, sounds, and scores, shedding light on their mutual desire for community and belief in sound as power. The audience was then invited to join in the discussion in 'a collaborative audio, visual, and spoken conversation'.[14] Lacy's *Silver Action* enabled hundreds of feminists over 60 to tell protest stories at patterned tables at the Tate Modern, scribed by young men to be projected in real time on gallery walls and tweeted out by young women.[15] The obvious synergy between these celebrations of gendered and feminist voice and memory is why we should count these artworks alongside more obviously historical projects such as the interviews of Jill Liddington, Jill Norris, and Sherna Gluck with elderly suffragettes and suffragists in the early 1970s and why, indeed, it was thrilling that *Sisterhood and After* interviewees took part in Lacy's action. Although obviously there are differences in short, choreographed, public storytelling and the long form of oral history, the former reveals the latent performativity in the latter and demonstrates where the soundscape is itself meaningful as historical record, community, and aesthetic relationship.

Admittedly, as we leave behind the straightening terms of document-based, positivist historiography, the questions that feminist oral historians have asked ourselves about 'voice' as conduit for a collective consciousness become more pressing. Steve Connor argues that voice is crucial to the process whereby, in modern societies, political questions have come to be focused in the question of the subject. But, like the 'subject', voice has been deconstructed. As Sherna Gluck and Daphne Patai put it, '[T]he interview is a linguistic, as well as a social and psychological, event', one that requires much more from a feminist interviewer than 'displaying empathy', and which may often result in discovering painful differences between women. On one level, this provokes feminists to 'learn to listen' more responsibly, sometimes even borrowing from therapeutic techniques to do so (Anderson and Jack). However, a sound art perspective asks us to think in addition about the interview 'event' not only as grounds for linguistic or political negotiation but as body-event, soundscape, first-draft performance. Through considering oral history's natural relationship to both music and theatre, we can acknowledge fully the 'peculiarities' of oral history as both form and relationship, its obviously creative as well as deconstructive relationship to both the past and to the audience (Portelli). Unlike a reading public constituted of individuals in isolation, a listening public is more likely to be made up of listeners inhabiting a condition of plurality and intersubjectivity, one that involves pleasure as well as challenge in both listening and performance (Lacey 8). Pauline Oliveros, an outstandingly successful feminist sound artist, provides an endpoint for oral historians to consider in this respect. Her minimalist compositions propose that sharing sound can bring us into transcendental community, if we listen hard enough not only to each other but also to the environment that connects us. These compositions are choral not because they involve voices but rather improvised responses by instrumentalists or sound makers of all kinds, in which no one musician can lead for long (Mockus). She calls this 'deep listening' (Oliveros).

Broadcast radio, of course, has also long provided a model for how to orchestrate interviews in a richly textured audioscape, as well as a home for recorded life narratives in many forms (McHugh). Again, though, it is where radio converges with sound artistry that we can see the greatest inspiration for oral historians. From classics such as Charles Parker's, Ewan McColl's, and Peggy Seeger's BBC *Radio Ballads* of the 1950s to the ongoing 'Listening Project', it is the grain of the interviewee's voice, its cadences and cracks, in the unsettling background noises, the inadvertent interruptions, which may tell us most about another's life-world. Ken Cormier indeed compares this awkward sonic environment to the emotional 'punctum', or touching personal detail, that hides within the 'studium' of social-historical narrative (Cormier 423). Digitisation has further complicated any romantic ideal of the unmediated authentic oral historical voice, but similarly has made it far easier to preserve its orality with obviously creative invitations. Now the oral historian, the broadcaster, the archivist, the curator, the sound engineer, and the sound artist may switch roles. This can be disconcerting for an audience used to a particular genre or polished production quality, for here the 'raw' and the 'cooked' often get served simultaneously. Yet as so many have observed, the cross-platforming ability of the technology has brought with it a democratisation of memory, as well as wonderful new opportunities for experimentation with everyday sounds (Hardy). Family stories, which remain at the heart of the digital biographical repertoire, are ripe, then, for re-sounding (Hardey).

Early in the project, we had already begun to digitally 'cook' the *Sisterhood and After* oral history, through creating a richly illustrated Learning Website at the British Library, where 125 clips from the interviews are curated through 10 narratives of movement activism—'Families and Children' was notably the first —with questions and exercises for schoolchildren. Barbara Jones as a builder was an obvious subject for one of the 10 additional short videos, and Lizzie Thynne, the project filmmaker, happily pictured Barbara knocking nails into a slate roof and demonstrating lime plastering to young women.[16] However, this film, as with the website, was designed to be both realist and didactic. It was a great pleasure, then, to work with Thynne in producing *Voices in Movement*. In its first form, this was a multichannel sound installation at *Public and Private Archives: Creative Negotiations*, at Sussex University, April 2014.

It was then shown as a single-screen work as part of the group show *Family Ties; Reframing Memory* at the Peltz Gallery, London, 3–25 July 2014. This 15-minute soundwork has pushed the 'postdocumentary' potential of our recordings to complicate any direct correspondence between voice, person, and location. Instead, Thynne denaturalises and plays with the vocal and the testimonial, drawing out the formal and relational elements of the interview as constructive space. The gallery context is more conducive to this kind of experiment than other forms of exhibition. Thynne uses visuals sparingly—such as fleeting archival snippets of period home movies and only briefly glimpsed photographs of the speakers, thus providing a meditative texture quite different to the more typical kinds of oral history clips exhibited in state-funded community oral history projects, including the kind of realist voice that we had already curated on our website.

Figure 1. *Voices in Movement* sound installation, Sussex University, April 2014. Photo: Alexandra Loske

Obviously we are still welcoming oral history's capacity to convey a feminist vision through experiential life history. Further, Thynne clearly draws on the family story to provoke emotional connection and understanding, to attempt to find, in other words, the 'punctum' in the oral histories she had selected. Her design of the sound installation has primarily allowed a different way to represent that struggle. By evoking family relationships in a more oblique, layered, or fragmented form, she aims to capture a psychic realism that mimics the process of memory itself—contested, sometimes traumatised, but also necessary. Thynne comments,

> The voices of Taylor and Jones are intermittently counterpointed with the found home movie footage, from the 1940s. The blurry and fleeting appearance of the

Figure 2. *Family Ties: Reframing Memory Exhibition*, Pettz Gallery, London, July 2014. Photo: Sally Waterman

Figure 3. 'But you see there was something going on for her, because she was already pregnant', *Voices in Movement*, Peltz Gallery, 2014. Photo: Lizzie Thynne

> latter, between sections of black screen, shows some of the stereotyped images of families from the genre—little girls playing in the sunshine, proud mothers and fathers presenting new babies to the camera. The occasional appearances of these images refuses to anchor the oral interview fragments, mimicking the process of memory itself and underlining how the voices themselves are detached in time and place from the visual scenes of 'family life'. (Thynne)

The cutting up and repetition of words that Barbara Taylor uses in her already insistently repetitive speech, I believe, enhances the already rhetorical qualities of her interview. In the darkened room, the screen teasingly remains darker, and the voices come and go, panned between speakers. The first is mine, to show that this is not a neutral account:

> [Margaretta Jolly][...] to also acknowledge what you said previously about, the particularity of the kinds of memories that you have.
> [Barbara Taylor] Yes.
> [MJ] And, that this isn't necessarily the only story, and so, of course, say other things if you want as they come along.
> [BT] Yes yes. Yes.
> [MJ] But you know, also to acknowledge that there's a, a darkness to what you've said.
> [BT] Yup. Yup. Yup, yup. (excerpt from Thynne and Hughes)

Further, Thynne selected these clips from the two Barbaras to show family as a site of isolation and shame for girls and mothers, but one that can be transformed through politicisation and the revelation of secrets.

> I definitely felt if I told anybody the whole family would fall apart. My role was to keep the secret and be responsible. (Barbara Jones excerpt from Thynne and Hughes)

This situation, of course, creates a plot of discovery that carries the listener through when there is little to look at. Constructing this initially as a paper edit alongside a rough cut of the audio clips placed on a timeline, Thynne built upon this basic structure to design an oral journey of collective understanding. We segue from Jones's recollections of her family's secret—her abuse as a child—to a montage of interview clips from other activists. The activists reflect on how their involvement in feminism changed their views of the family and of gender within it. These reflections, occasionally juxtaposed with bursts from Sue Crockford's iconic film of women chanting on the first WLM march, highlight the deep personal changes that accompanied this more public face of protest and political visibility.[17] Thynne comments,

> In the first half of the work, we hear the individual interwoven stories of Taylor and Jones which suggest how they have revised and re-contextualised their recollections of childhood in the light of their politicisation; in the second half, the montage of brief clips from other women acts like a chorus showing how the women's movement gave them the political tools to understand the imprisoning power relations in their own families. I utilise a fragmented editing style, where sometimes one speaker even completes the sentence of another to suggest the

collective political consciousness which underpins their analysis of their histories. (Thynne)

We hear recovery, humour, and forgiveness as well as critique. Cynthia Cockburn comments on the effect of her newly acquired feminist understanding on her view of her father: '[I]t allowed me to place him in the patriarchal system [...] understanding the incredibly impoverished circumstances in which he was living', while Kirsten Hearn jokes that if she'd remained heterosexual, she'd have probably been 'well heeled' by now. The final voice is Nadira Mirza's, as she describes seeing a young teenage mother with a pushchair today, wondering what choice she has really made. Her musing tone leaves the question open for the listener to answer (The British Library).[18]

This vocal fugue is vitally enhanced by music and sound effects composed by contemporary classical musician Ed Hughes. Playing with these in draft form, we realised how easy it is to tip from the minor key into unfortunate horror film suspense or, conversely, too literal comic 'illustration'. Hughes provides instead the delicately ironic sound of a crowd of children cheering when Taylor says, 'Looking at me in photographs, you don't know if you are looking at a little girl or a little boy', of smashing glass when she describes dropping a bottle of milk, and a 'signature' sound of a climbing and descending piano drone (excerpt from Thynne and Hughes). Barbara Jones's matter-of-fact tones as she describes the family church routine alongside dreadful abuse is juxtaposed with quiet altar bells and disconcerting tingling and scratching noises, a sound also heard later when Jones recalls the story of her mother's pregnancy. Oddly, Hughes's beautiful string quartet compositions proved too like voices themselves and therefore competed with the oral history recordings, and so they were reluctantly removed. Hughes commented on how the intricate structure of music can often overwhelm other elements of sound and visual structure, which is why it often has to be written fresh for specific films. However, he suggested using an oboe solo for the closing credits. This allowed us to find a space for music that the complex fugues of voices in the body of the piece could not permit, but whose rising musical idea was important to leave with the listener as encouragement to take up the piece's challenges.

Hughes also advised on how to manipulate the spatial as well as acoustic dimensions of sound installations. The first iteration of *Voices in Movement* distributed tracks to different speakers, panning the voices to the front of the gallery and the music and effects to the back, with the aim of surrounding the listener, creating a polyphonic, serious, and concentrated space, enhanced by the darkened room. When shown at the Peltz Gallery, the work was re-versioned into a stereo, as opposed to multichannel, piece, and displayed on a monitor with the sound on headphones. This was due to the space constraints as it was exhibited there as part of a group show. Thynne comments,

> Since visitors' first encounter with the work at the Peltz was through the sight of the monitor screen with no audible sound, this necessarily placed more emphasis on addressing visitors through the screen as opposed to through sound and

additional visual ideas were introduced. These included using found footage with a metaphoric relationship to the stories being told, adding a new dimension to the original sound/image correlations. Jones tells of how her mother spent the night in a cinema trying to decide whether to marry her father because it would mean taking the big step of converting to Catholicism. The cinema reference is taken up in the image track showing a mid shot of a woman on a scratched segment of film. The woman is looking disconsolately at the scratch lines apparently hemming her in, which are actually on the surface of the film and

Figure 4. *Voices in Movement*, Peltz Gallery, July 2014. Photo: Alexandra Loske

would, of course, not usually be visible to a protagonist but which here suggest the mother's entrapment. A following shot shows the film strip of the woman with damaged sprocket holes, again indirectly evoking the mother's situation, finding herself pregnant before marriage in the 1950s. These clips are a serendipitous find from a 1950s training film about how to project a film which here surreally works to comment on the emotional undercurrents of Jones's story of her mother's marriage rather than to illustrate it. (Thynne)

In my attempt to summarise weeks of discussion and creative listening, I realise that I risk abstracting and trivialising the sounds of very personal recordings, unable of course, in the print form of this article, to fully convey their effect.[19] Evidently an aesthetic approach does not exempt us from the ethical challenges of life narrative, though it should be said that the interviewees were asked and gave permission for the oral history's use in this form. Yet if we consider the responses by listeners who heard it at the Peltz Gallery's *Family Ties* exhibition, we find that the arrangement of sounds and voices in the piece provoked new thoughts about the meaning of these women's family experiences.[20] One listener, who felt that 'the darkness [between stretches of film archive footage] allows a looking away', commented that

> the voices and footage say as much about [...] the silences and things not seen [...] as they do about what is actually heard and seen. In regard to memory and family function/dysfunction, this seems entirely appropriate. The way the piece is put together leads the viewer/listener to a deep empathy with the children

Figure 5. *Voices in Movement*, Peltz Gallery, July 2014. Photo: Alexandra Loske

portrayed, partly because we 'know' that these children are no longer children, and yet somehow we know they co-exist with the adults speaking. (Clare Best, qtd. in Loske)

Another commented that the piece had a 'Proustian' effect on him, taking him back to 'an earlier self', stirring emotional responses. This listener also noted that 'what was interesting to me was to see how the concept of family changes over time':

> What I remember was that the family was the enemy of human liberation; family was a prison from which you had to break out, [whereas] today families are seen as a place of refuge, what to go back to from the nasty world outside, and it just wasn't like that in the 1970s. Families were the nasty, evil people from where you want to get away. [...] just watching *Voices in Movement* for 15 seconds, that's what I was feeling. (Simon Harker, qtd. in Loske)

What he hears suggests a sobering loss of feminist critique in today's conservative family ideology, and the possible function of sound pieces as collective political memory alongside the more obvious oral history archive. On the other hand, we see the contingency of listening, and the difficulty, even in sound art, of representing the plurality of lives and feminisms, too.

Conclusion

A 15-minute sound piece cannot, of course, encompass the complexity of the individual narratives of our oral history interviewees. Even in this elliptical and aesthetically demanding form, it is tempting to use well-known plots rather than, for example, acknowledging that feminist mothers could also be disappointing, or that blood family has returned for many as the perceived 'refuge' in ever more unsettled times. But *Voices in Movement* has been an important development for our oral history project. Thynne and Hughes open up chronological life narrative recordings in ways that provoke a different kind of reflection, critical and emotional at once, about the on-going search for the postpatriarchal family. At the edges of narrative and visual form, this also entails a different form of feminist witnessing and 'deep listening', one that allows for the inchoate and hopeful qualities of music. Looking back, these activists show a new melding of biological with affective ties, and they tell of chosen families and sustaining friendships across new generations. This is as much the 'movement' that the sound piece tries to capture as any escape from the oppressive structures of the mid-twentieth-century families they knew. These must surely be the sounds of feminism beyond the archive and the gallery, ones that I hope we may hear anew in everyday life.

Notes

[1] Examples include Lesley Abdela's story of her businessman father's care for her when her mother died in her early teens, or Jalna Hanmer's sense that she inherited her left-leaning father's burdens. See *Sisterhood and After: The Women's Liberation Oral History Project*, Catalogue reference C1420/13 and C1420/04. © University of Sussex and the British Library.

[2] Beatrix Campbell, interviewed by Margaretta Jolly, *Sisterhood and After: The Women's Liberation Oral History Project*, Catalogue reference C1420/60. © University of Sussex and the British Library.

[3] Our interview with Rowena Arshad is a case in point. Arshad's mother had bravely brought her daughter up alone, after being abandoned by her husband. Investing her hard-won savings to get her daughter to boarding school in England, she eventually followed her daughter to settle near her, despite racism and loneliness. Yet these maternal sacrifices were accompanied by judgement, control, and racism of her own. Arshad's later professional and political success seems to have been at least enabled by, if not directed towards, a very solid and happy family life of her own,. in which she found also more kindred spirits in her husband's Scottish activist parents. At the same time, she talks about remaking rather than rejecting traditions of childhood obedience, nurturing a social conscience that her own mother lacked. Rowena Arshad, interviewed by Rachel Cohen, *Sisterhood and After: The Women's Liberation Oral History Project*, 2010–2013. Catalogue reference C1420/21. © University of Sussex and the British Library.

[4] Barbara Taylor, interviewed by Margaretta Jolly, *Sisterhood and After: The Women's Liberation Oral History Project*, 2010–2013. British Library Sound & Moving Image Catalogue reference C1420/38, transcript p. 205; Track 8. © University of Sussex and the British Library.

[5] Broadcast 17–21 February 2014. http://www.bbc.co.uk/programmes/b03vd5j7.

[6] Taylor transcript p. 197; Track 8.

[7] Taylor transcript p. 29; Track 1.

[8] Office of National Statistics, 'EMP14: Employees and Self-Employed by Industry'. 12 November 2014; http://www.ons.gov.uk/ons/publications/re-reference-tables.html?newquery=*&newoffset=25&pageSize=25&edition=tcm%3A77-331783. See also http://www.theguardian.com/women-in-leadership/2014/jul/23/why-bob-the-builder-is-keeping-women-out-of-construction.

[9] Barbara Jones, interviewed by Margaretta Jolly, *Sisterhood and After: The Women's Liberation Oral History Project*, 2010–2013. British Library Sound & Moving Image Catalogue reference C1420/53, transcript p. 69, Track 3. © University of Sussex and the British Library.

[10] Jones transcript p. 111, Track 4.

[11] Jones transcript p. 1, Track 1.

[12] Jones transcript p. 29, Track 1.

[13] All interviews available in the *Sisterhood and After* collection at the British Library, as above, including short interviews with the core team, including Jolly and Thynne as part of the project's documentation.

[14] http://flattimeho.org.uk/events/someone-else-can-clean-mess-annea/.

[15] http://www.suzannelacy.com/silver-action-2013/.

[16] See bl.uk/sisterhood.

[17] *From a Woman's Place* (dir. Crockford, 1971).

[18] All interviews available in the *Sisterhood and After* collection at the British Library; cited here as in Thynne and Hughes.

[19] The screen soundpiece version of *Voices in Movement* is viewable at http://vimeo.com/100939494.
[20] http://www.bbk.ac.uk/arts/research/peltz-gallery/past-events-and-exhibitions-at-the-peltz-gallery/3-25-july-2014-family-ties-reframing-memory.

References

Amos, Valerie, and Pratibha Parmar. 'Challenging Imperial Feminism'. *Feminist Review*. 80 (2005): 44–63.
Anderson, Kathryn, and Dana C. Jack. 'Learning to Listen: Interview Techniques and Analyses'. *Women's Words: The Feminist Practice of Oral History*. Ed. Sherna Berger Gluck and Daphne Patai. New York: Routledge, 1991. 11–26.
Banks, Olive. *Becoming a Feminist: The Social Origins of 'First Wave' Feminism*. Brighton: Wheatsheaf, 1986.
Barrett, Michèle, and Mary McIntosh. *The Anti-Social Family*. 1982. 2nd ed. London: Verso, 1991.
Barrett, Michèle, and Mary McIntosh. 'Ethnocentrism and Socialist-Feminist Theory'. *Feminist Review* 80 (2005): 64–86.
Bechdel, Alison. *Fun Home: A Family Tragicomic*. London: Jonathan Cape, 2006.
The British Library. *Sisterhood and After: The Women's Liberation Oral History Project*. 2010–2013. Sound & Moving Image Catalogue C1420. © University of Sussex and the British Library. <http://cadensa.bl.uk/uhtbin/cgisirsi/?ps=dyEqWB8Y9A/WORKS-FILE/275090033/18/X087/XNUMBERS/C1420>
Buss, Helen M. *Repossessing the World: Reading Memoirs by Contemporary Women*. Life Writing Series [Canada]: Wilfrid Laurier UP, 2002.
Butler, Anne M., and Gerri W. Sorenson. 'Patching the Past: Students and Oral History'. *Women's Oral History: The Frontiers Reader*. 2002. Ed. Susan Armitage, Patricia Hart, and Karen Weathermon. Lincoln: U of Nebraska P, 2003. 196–210.
Chamberlain, Mary. *Fenwomen: A Portrait of Women in an English Village*. London: Quartet, 1975.
Connor, Steven. *Dumbstruck: A Cultural History of Ventriloquism*. Oxford: Oxford UP, 2000.
Cormier, Ken. 'Writing the Tape-Recorded Life'. *a/b: Auto/Biography Studies* 27.2 (2012): 402–26.
Cosslett, Tess, Celia Lury, and Penny Summerfield. *Feminism and Autobiography: Texts, Theories, Methods*. Transformations. London: Routledge, 2000.
Couser, G. Thomas. *Memoir: An Introduction*. New York: Oxford UP, 2011.
Gluck, Sherna Berger, and Daphne Patai. *Women's Words: The Feminist Practice of Oral History*. New York: Routledge, 1991.
Hardey, Michael. 'Digital Life Stories: Auto/Biography in the Information Age'. *Auto/Biography* 12.3 (2004): 183–200.
Hardy, Charles, III. 'Authoring in Sound: Aural History, Radio, and the Digital Revolution'. *The Oral History Reader*. Ed. Robert Perks and Alistair Thomson. 2nd ed. London: Routledge, 2006. 393–405.
Hirsch, Marianne. *The Mother/Daughter Plot: Narrative, Psychoanalysis, Feminism*. A Midland Book; Mb 532. Bloomington: Indiana UP, 1989.
Jolly, Margaretta, Polly Russell, and Rachel Cohen. 'Sisterhood and After: Individualism, Ethics and an Oral History of the Women's Liberation Movement'. *Social Movement Studies* (2012): 1–16.
Lacey, Kate. *Listening Publics: The Politics and Experience of Listening in the Media Age*. Cambridge: Polity, 2013.

Lacy, Suzanne, Moira Roth, and Kerstin Mey. *Leaving Art: Writings on Performance, Politics, and Publics, 1974–2007*. Durham, NC: Duke UP, 2010.

Loske, Alexandra. 'Gauging the Impact of an Oral History-Based Sound Project: Voices in Movement'. Unpublished Report. University of Sussex. 4 October 2014.

Lynch, Claire. 'Who Do You Think You Are? Intimate Pasts Made Public'. *Biography* 34.1 (2011): 108–18.

Marchant, Alison. 'Treading the Traces of Discarded History: Oral History Installations'. *Women's Oral History: The Frontiers Reader*. 2002. Ed. Susan Armitage, Patricia Hart, and Karen Weathermon. Lincoln: U of Nebraska P, 2003. 183–95.

McHugh, Siobhán. 'The Affective Power of Sound: Oral History on Radio'. *Oral History Review* 39 (2012): 187–206.

Miller, Nancy K. *Bequest & Betrayal: Memoirs of a Parent's Death*. New York: Oxford UP, 1996.

Mockus, Martha. *Sounding Out: Pauline Oliveros and Lesbian Musicality*. New York: Routledge, 2008.

Oakley, Ann. 'Interviewing Women: A Contradiction in Terms'. *Turning Points in Qualitative Research: Tying Knots in a Handkerchief*. 1981. Ed. Yvonna S. Lincoln and Norman K. Denzin. Walnut Creek, CA: AltaMira P, 2003. 243–64.

Oliveros, Pauline. *Deep Listening: A Composer's Sound Practice*. New York: Universe, 2005.

Perks, Robert. *Review of the Year*. London: British Library, 2014.

Portelli, Alessandro. 'What Makes Oral History Different'. *The Oral History Reader*. 1979. Ed. Robert Perks and Alistair Thomson. 2nd ed. London: Routledge, 2006. 32–42.

Stanley, Liz. *Feminist Praxis: Research, Theory, and Epistemology in Feminist Sociology*. London: Routledge, 1990.

Taylor, Barbara. *The Last Asylum: A Memoir of Madness in Our Times*. Hamish Hamilton, 2014.

Thompson, Paul Richard. *The Voice of the Past: Oral History*. 3rd ed. Oxford: Oxford UP, 2000.

Thynne, Lizzie. Personal Communication. December 2014.

Thynne, Lizzie, and Ed Hughes. *Voices in Movement*. 2014, Peltz Gallery, London.

Vaccaro, Annemarie. 'Toward Inclusivity in Family Narratives: Counter-Stories from Queer Multi-Parent Families'. *Journal of GLBT Family Studies* 6.4 (2010): 425–46.

Yow, Valerie Raleigh. *Recording Oral History: A Guide for the Humanities and Social Sciences*. 2nd ed. Walnut Creek, CA: AltaMira, 2005.

Yow, Valerie Raleigh. 'Textile Women: Three Generations in the Mill'. *Southern Exposure* 4.3 (1976): unpaginated.

The Epistolary Dynamics of Sisterhood Across the Iron Curtain

Leena Kurvet-Käosaar

The article focuses on the correspondence between two Estonian sisters (the author's maternal grandmother and great aunt) across the Iron Curtain from 1956–89, offering an analysis of the capacity of the epistolary medium in maintaining an intimate bond between the sisters. Although deeply personal in nature, the dynamic of the correspondence is nevertheless shaped by larger historical and sociocultural forces, including censorship. The analysis traces development of different strategies of intimacy, such as reliance on common memories, verbal manifestations of closeness, the importance of blood relations and familiarising each other with the details of everyday life.

Your letter arrived on 20 Sept [1956]. What a wonderful surprise: all these long years of silence and now, at last, your letter. As if you suddenly are again among us [and] I am as if talking to you across all these years. [...] Twelve years ago you were suddenly lost in the turmoil of war. We thought that you had perished in the war. [...] I sit here and write as the midnight is drawing near and a beautiful symphony is playing on the radio. No words can describe how I feel. What can I write in just one letter? I should write a book, not a letter. (Helga Sitska, 1 October 1956)

This letter marks the beginning of the correspondence between my maternal grandmother, Helga Sitska (née Valgerist, 1914–1989), who lived all her life in Tartu, Estonia, and her younger sister, Aino Pargas (née Valgerist, 1923–2000), who fled Estonia during the mass emigration in the fall of 1944, settled down in the United Kingdom after the war, and in the late 1950s moved to the United States. The correspondence between the sisters lasted for more than 33 years, until the death of my grandmother in April 1989. As travel into and from the countries behind the Iron Curtain was severely restricted during their lifetime, the sisters were able to meet only a few times. Letters became their only medium of communication. Their lives, on two continents and in vastly different socio-political and material contexts, were not only represented through the correspondence, but in a way 'lived' within the possibilities and boundaries of the medium.

Their correspondence is not exceptional, in that epistolary exchanges between family members, friends, and intellectuals constituted an important form of communication within Baltic communities separated by the Iron Curtain, and there have been numerous large-scale projects carried out in Estonia as well as in other Baltic states for the collection of life writings from this period. The focus of these projects has primarily been, however, on memories of the Soviet occupation and mass emigration to the West, and less attention has been paid to forms of communication and genres of life writing that connect the two worlds and breach the Iron Curtain.[1]

While some examples of correspondence across the Iron Curtain between Estonian intellectuals and literary figures are preserved in the Estonian Literary Museum, almost no resources of a similar nature are available for research in relation to epistolary exchanges between family members. Therefore, the correspondence of Helga Sitska and Aino Pargas, consisting of approximately 500 letters and running through nearly four decades, offers a unique insight into the sociocultural dynamics of the Cold War period and problematises the assumption of the impermeability of the Iron Curtain on an everyday level. In this article I discuss the possibilities and limits of the epistolary medium in maintaining contact between family members and, in particular, the role of letter writing as a vehicle for the intimate dynamics of sisterly affection.

Belonging to the category of 'everyday letter writing' (Barton and Hall 2–3) or '"letters" en masse' (Jolly and Stanley 93), the correspondence offers insights into social contexts and practices of letter writing rather than exploring 'the relationship between "life" and "art" referentially inscribed or implied' in letters by literary figures that has been of primary interest in academic research on letters (Jolly and Stanley 94). However, as Margaretta Jolly has pointed out, it is important not to overlook 'the rhetorical dimension of even familiar letters' traceable, for example, via the tone or, according to some scholars of letters, the '"epistolary performance" or "personae"' adopted in the letters' (Jolly and Stanley 93; see also Cockin 152; Maybin 152). Kathleen A. DeHaan views epistolary performance 'as a contextual discourse, which appears in the negotiation between the participants' (108) that is closely related to the process of construction of identity, in particular with regard to life experience of a deeply transformative nature, such as, for example, immigration.

The correspondence between Helga and Aino makes visible various agendas and influences (sociopolitical, family-oriented, national, gendered, intimate) that shape the epistolary exchanges and ultimately the nature of the relationship itself. To a large extent, these factors only come to bear on the correspondence through the textual, poetic, and rhetorical means employed. The dialogical aspect of letter writing, which Liz Stanley posits as one of the key features of epistolarity (202), can be traced in the correspondence with exceptional clarity and detail. The letters have the quality that Liz Stanley refers to as 'a life of their own' (210); conversely, the sisters can be viewed as having a life of their own in letters or through the letter-writing practice that, as the letters themselves clearly make visible, forms an important and unique part of their life experience.

A Bundle of Letters

The significance that the sisters attributed to each other's letters made them preserve them with loving care, thus granting the letters 'an afterlife', an aspect of this form of correspondence that, according to Jolly, opens up a range of different questions: 'what happens to personal letters after they are written and received is itself a significant psychological and ethical as well as literary issue' (206). For me, working with the correspondence that is part of my family's archive has been a complex process of negotiation between a number of different roles that the task has required—most important, that of a family member assuming the role of a family archivist and that of a researcher mediating aspects of the correspondence of interest from the point of view of scholarly research on letters. I have also realised that even in the case of a collection of correspondence in private ownership, the reasons for its preservation form an important starting point for both its scholarly and private, family history-related consideration.

The correspondence, both sides of which have been preserved, constitutes the most extensive textual record of my family's history ever produced. This fact is perhaps not surprising, as the letters can be viewed as belonging to the category of immigrant records that, according to Elliot, Gerber, and Sinke, are considered to form 'the largest body of the writings of ordinary people of the past that historians and other researchers possess' (3). The preservation of both sides of the correspondence in immigrant records of this kind has been attributed to the generally high professional and educational status and residential stability of the families concerned, as well as the symbolic value of the letters as proof of establishing a branch of the family in a foreign country (Helbich and Kamphoefner 30; Gerber 47). Some of these criteria apply to my family, as both sisters were relatively well educated, and in Estonia my family has lived in the same home for over 70 years. Yet Aino, in spite of moving from the United Kingdom to the United States and within the United States, also kept all her sister's letters, and my mother brought them back to Estonia when she visited her aunt shortly before her death in 2000. My grandmother preserved all the letters and sorted them year by year into bundles that she tied with ribbons of different colours. By the time I systematically started reading the letters, the sisters had already well organised both sides of the correspondence. There was, however, a bundle of letters among each half of the correspondence that did not follow the chronological order. Unfortunately, in my ignorance I rearranged the letters before I was able to grasp the significance of their separation from the annual collections. Although it is impossible for me to restore these packages now, I have realised that they probably contained letters that the sisters would read again and again, letters that for them provided the strongest proof of each other's existence and emotional connection and were of most support during difficult times.

Helga and Aino's correspondence, in my opinion, was first and foremost preserved because of the highly affective and intimate value of the letters for them, offering insights into what Toni Morrison has referred to as 'emotional memory', foregrounding those details of experience that are affective, sensory, often highly specific, and personal (Morrison 99; Cvetkovich 242). This also relates to the materiality of the letter form, its 'psychological power [...] as a form of bodily trace that underwrites, and sometimes dominates, its text' (Jolly 208; see also Cook). As much as the correspondence functioned for the sisters as textual proof of each other's existence, just as important was the simple physical materiality of 'a bundle of letters'. David Gerber has pointed out that 'immigrant letters are not principally about documenting the world, but instead about reconfiguring a personal relationship rendered vulnerable by long-distance, long-term separation' (143). The lifelong correspondence across the Iron Curtain of my maternal grandmother with her younger sister was primarily about developing and sustaining an epistolary bond, an intimacy in letters that functioned as a substitute for face-to-face relationship. Careful preservation, reading, and rereading of each other's letters, as well as addressing that in the correspondence, were an important part of that relationship.

Martha Hanna, in her research of the correspondence of frontline French soldiers with their family members during World War I, comes to a similar conclusion, arguing that 'to understand why frontline soldiers cherished letters from wives and parents as sacred objects, and why civilians preserved letters from the front with equal reverence, one must look beyond the testimonial content of wartime letters and analyse their affective and emotional functions and implications' (1342). Although deeply private and personal in nature, the epistolary relationship of the sisters is shaped by larger historical and socio-cultural forces that, in important ways, have shaped its tone (longing) and choice of topics (the need to update each other with daily life and avoidance of politically risky material), as well as its duration and frequency.

Writing in Code

Helga and Aino grew up in Tartu and attended the first Estonian-language gymnasium for girls in that city. Before World War II, Helga started studying law at Tartu University and graduated during the time of the German occupation of Estonia. She married while still a student and had two sons (born in 1938 and 1945) and a daughter (my mother, born in 1943). Helga lived in Tartu in the same apartment (which was also my childhood home) all her life. For many years she was considered unemployable by the Soviet regime for political reasons but finally found work in the field of the Soviet equivalent of real estate law. Aino left Estonia in the fall of 1944, and by 1956 had been living in Harlow in the

United Kingdom, where, according to her letters, she was quite happy with her life. After reuniting with her husband, who had been a prisoner of war in the Soviet Union for six years, she moved to the United States in 1958 and eventually settled down in Maryland. Aino and her husband did not have children. For more than 30 years, Aino worked at the Library of Congress in Washington, DC.

The extract from Helga's letter that opens this article seems to come from her very first letter to her sister.[2] There is, however, yet another letter, written in August 1956, that was smuggled to Finland and posted from there. Helga likely received her sister's address in a letter that found its way to Estonia in a similar manner. In her letter, Helga writes, 'Why I haven't written, my dear, needs no explanation in my opinion', and then tells her sister, 'When writing to me, make the letter look as if you were the first to start a correspondence' (21 August 1956). Only a few years before, those with relatives abroad who had left Estonia for political reasons were at risk of persecution, one of the reasons for the wave of deportations in 1949. Thus, this may have been the 'obvious reason' for not writing that Helga mentions in her letter posted in Finland. The complex, multi-layered beginning of the correspondence, which I was only able to work out after consulting with my mother, makes visible the intricate manner in which it needed to be navigated in order to get past the Soviet censorship system and avoid harming the people concerned.

In her analysis of the possible effects of censorship on the correspondence of frontline soldiers to their families, Hanna argues that the writers 'established codes and subterfuges to thwart the censors and nullify censorship's numbing effect' (1339). Although Soviet censorship was very different from that applied in France during World War I, the strategies of ordinary people in handling the potential censoring of their correspondence are likely to have similarities, regardless of the political regime or time period. In Helga and Aino's letters, examples can be found of both strategies: the adoption of what can be referred to as codes with regard to certain subjects as well as subterfuges—ways of handling some topics in an ambivalent and possibly deceptive manner.

An example of the use of a code is visible in Helga's response to Aino's inquiry about Tiiu, a good friend of Helga's. '[She] lives somewhere far away, somewhere in the Soviet Union', Helga writes, continuing, '[Q]uite a few of those who also lived in other, more faraway places have returned over the past few years' (2 February 1958). Helga's response seems to imply that she might no longer be in touch with her friend on a regular basis. In reality, Tiiu, who remained Helga's best friend throughout her life, spent more than 25 years in the Gulag, a fact that Helga could convey to Aino in her letter only in code. In Helga's letter, reference to returning (to Estonia) implies the amnesty that was granted to many camp inmates and deportees during the years following Stalin's death in 1953. Tiiu was not among those and could not return until the late 1960s.

An example of a possible effort to evade the attention of a censor by way of subterfuge appears in Helga's letter from December 1956. In response to Aino asking whether she should send them some food products, such as sugar, coffee, or chocolate, Helga responds, 'Dear Aino, please do not send me food products.

Together with Eedi [her husband, my grandfather Eduard Sitska] I can cater for my family's needs. Believe me; my family is properly fed although I spend quite a bit of money on that. As you can see, I am still a most inefficient homemaker' (12 December 1956). For a reader who does not know Helga personally, in this section she merely seems to highlight her poor skills of managing everyday life. However, by the time that Aino left Estonia, she must have been quite familiar with Helga's dedication to her family and her resourcefulness in solving various emergency situations caused by the radical changes that the war brought about. Helga's subsequent letters indicate that she was the main provider and carer for her family, financially as well as with regard to different household chores, and that she managed her responsibilities quite efficiently. Although Helga's self-assessment can be attributed to modesty, it can also be read as a covert way of informing her sister about the quality of everyday life in Soviet Estonia as well as signalling that she was unable to write openly about these matters. Other similar instances can be found in the letters that show that both corresponding parties were aware of the possibility of censorship, and in order to avoid a run-in with it, the sisters made a sustained effort to address politically risky topics in a covert and ambivalent manner. At the same time, nothing in the text of the letters or on the envelopes bears any identifiable marks of censorship, such as, for example, crossed-out words or markings on the envelopes.

Strategies of Intimacy

The letters, however, also show that different layers of meaning can exist side by side, successfully serving multiple purposes. For the possible censor, Helga's first letter convincingly expresses her surprise upon receiving a letter from her sister after 12 years of silence. It reveals her happiness upon re-establishing contact with Aino after years of worry and lack of information about her fate. Recognising this, Aino is deeply moved by the letter. 'I received your letter about a week ago', she writes on 22 October 1956. 'It is so hard to believe', she continues,

> that after so many years I am receiving a letter from home again. I look again and again at the photos of you and your children, and no words can express how dear you are to me. I think of you every night before going to sleep and feel as if I am together with you all.

As the quoted sections demonstrate, a strong desire to (re)establish and (re)cover an intimate bond between the sisters is apparent already in the very first letters. Over the years, their correspondence developed its own textual strategies for creating intimacy, and its evolution and transformation over time makes visible the ways in which the epistolary medium catered for the needs of maintaining and developing intimate exchanges at the same time as complying with externally imposed constraints. From my readings of mostly the letters of the 1950s and 1960s, three main strategies for creating intimacy emerge: reliance on shared memories, verbal confirmation of closeness, and various ways of familiarising each other with the details of everyday life.

The first of the three—reliance on shared memories—plays a particularly important role during the first years of the correspondence. 'I should ask both myself and you, do you still remember, do you remember?' Helga writes in February 1957, shortly after Aino's birthday. She continues,

> [M]emories can be painful but they are still a blessing to our soul. I remember well the day when you were born. I helped to give you your very first bath. Then we put you in your stroller, an old-fashioned one, with a huge rubber hood. I was asked not to push it very fast but in a gentle manner [...] and now so many years have gone by that I have pushed the strollers of my own kids past the toddler age' (10 February 1957).

It may be perhaps doubtful whether an eight-year-old girl was allowed to assist with giving a newborn child her first bath, and even more so if in February, the coldest winter month in Estonia, a newborn would have been taken out for a walk. However, from the point of view of the relationship dynamics, what counts is not the factual preciseness of the memory but its purpose and effect. By tying together taking care of her children and her sister, as if her sister was her child as well, Helga strives to create a maternal and caring image of herself. In the correspondence to follow, Helga assumes the role of the maternal older sister who looks after her little sister. Helga may have adopted this role because of the age difference of nearly nine years between the sisters, but it may also have been a way to alleviate Aino's feeling of loss and sadness on learning belatedly of the death of their parents in the early 1950s.

Aino responds by offering a memory from Penuja, my grandfather's farmstead, from which the family fled to escape the battles going on between the Soviet Army and the German Army in and around their hometown of Tartu in the late summer and early fall of 1944, when both sides bombed the city heavily. In Penuja, Aino departed from her family before she headed to the coast to flee from Estonia in September 1944, together with thousands of other refugees.[3] In November 1958 she writes,

> I read your letter from 28.10.58 again. You write so beautifully and your words go right to my heart and my soul yearns to be together with you again. [...] [A]gain and again I can see this last memory picture [from home], Mamma standing on the doorstep, holding little Mari with Andrus, a little guy of 6, by her side. When the carriage arrived to take me to the station, that's how they were left standing there, and an eternal memory [of this] was engraved in my heart for eternity. (14 November 1958)

This was the very last time that Aino saw her mother, Emilie—or Mamma, as she was called. In her memory she was holding my mother in her arms with her other grandchild, my uncle, by her side. Aino's memory focuses primarily on her mother but also includes Helga's children, and through this she brings her sister into her memory picture as well; in particular, as in the preceding sentence, she elaborates her great longing for Helga. In Aino's letter, her longing for her mother that cannot be satisfied, and her longing for her sister—perhaps even more intense because of the uncertainty about their future—blend into each other and tie her emotionally into her family. As she writes in an earlier letter, '[T]ime passes but memories live on, and often I tread in my thoughts the familiar paths that I once wandered along at home' (04 March 1957).

Sisterly Roles

The letters also indicate the importance of blood relations for both sisters; through blood they define themselves as the closest people to each other in the world. For example, in one of her letters, Aino, looking at Helga's photograph, comments, 'It seems to me we have similar features, we are also emotionally very close and understand each other well. It is really such a shame that we live so far from each other' (1 July 1957). Maarit Leskelä-Kärki, who has studied the construction of sisterly relations in the epistolary practices of the three siblings of the well-known Finnish (fennoman) family Krohn, argues that correspondence among siblings makes visible 'certain features of duty, force, and tradition; one is expected to write to siblings and to maintain family connections and traditions', as well as highlighting the importance of 'blood relation' as a feature of family relations at the beginning of the twentieth century (25, 27). Although Helga and Aino's correspondence covers a much later period, the sisters' perception of the nature and importance of family relations likely dates from their childhood.[4] As both lived their adult lives in circumstances dictated by the change of Estonia's political regime—one of them in exile, and the other in silent resistance to the regime—holding on to the values of their childhood and formative years remained an important foundation of their subjectivity.

The force of duty and tradition in maintaining family ties is expressed with different modality in the sisters' letters. Helga, as the eldest child in the family who took care of her parents during the last years of their lives, assumes the role of the family's head after their parents' death. Throughout their correspondence she reminds her sister that she belongs in Estonia with her family and expresses hope that someday she will be able to return. In the correspondence in general, references and responses to poems by well-known Estonian authors form a distinct strand, evoking a shared world not only in terms of immediate familial affiliations but also through a sense of belonging, along the lines of national identity. The sisters address several times in the letters one poem in particular: 'It Flies to the Hive' (1905), by the well-known Estonian poet Juhan Liiv, which can be regarded as having iconic status in Estonian culture. The poem draws a parallel between bees returning to their hive against storm and thunder ('thousands will fall on the way / still thousands will reach home') and the yearning of Estonians for their homeland and determination to return ('you forget death and suffering / and hurry toward the homeland' [93]). In September

1957 Aino writes to Helga, 'How often I have read this poem. [...] I yearn for peace, for rest and for a home! [...] [B]ut there is no home anywhere, one is and forever will be a stranger in a foreign land' (25 September 1957).

In her research on Estonians' life narratives of the journeys of escape to the West, Tiina Kirss highlights the relevance of their affective modalities, particularly feelings of loss, regret, and desire for consolation (616). Kirss points out that the shocking and traumatic effect of the journey 'needed time to be worked out and coped with that the refugees had neither in the neutral Sweden nor in Germany where the war was still going on' (616). Considering the political limits on the correspondence, it would not have been possible for Aino to write about her escape journey, yet her letters from the late 1950s and early 1960s make visible a perception of her past as a burden on her present.

In the very first letters, the sisters ask for each other's photograph. As neither of them owned a camera, it took some time for them to send recent photographs, and they were both very much looking forward to receiving them. The close attention they pay to the photographs confirms Patricia Holland's view of photographs as 'objects [...] which are part of our personal and collective past, part of the detailed and concrete existence with which we gain some control over our surroundings' (10). On the one hand, photographs are a way for the sisters to imagine a reality; both sisters mention how they always have a photograph of the other in front of them when writing a letter, so that they feel as though they are having a conversation with each other. On the other hand, photographs are also a source of doubts, suspicion, and longing. Both describe anxiously studying the images (in one letter, Helga even describes doing this with the help of a magnifying glass) and mediating to each other in detail their 'readings' of the photographs. Through the photographs they can, for example, trace possible changes in character over the years, as well as assess the compatibility of the information mediated by the photographs with that delivered in the letters. Photographs offer the sisters 'a framework within which [their] understandings of various realities can come into play' (Holland 4) that supports the bond created and maintained by the verbal part of the letters. On 9 November 1957 Helga writes to Aino,

> I'm looking at your photos over and over again. They are next to me when I'm writing to you, and I'm trying to imagine you doing different things. [...] My dear, when can I see you again? [...] [L]etters, letters—they are merely a substitute.

However, as the section quoted above demonstrates, photographs also often trigger a painful longing for a real-life face-to-face meeting. A similar reaction to photographs can be traced in Aino's letters:

> Many thanks for the photos. It breaks my heart to look at these lovely faces and read your letters. I have lived with this longing for a long time and will continue doing so. I hope, my dear Helga, that one day I'll be able to return to you. (19 November 56)

'I Should Write a Book'

Having been reunited through their correspondence after nearly a decade, the sisters eagerly ask for and send each other updates and overviews of their daily life and descriptions of their surroundings. For example, in one of her letters from the very first year of (re)establishing the correspondence, Aino writes about her life in Harlow, a town near London, as follows:

> You do not need to worry about me. I live quite well and am in good health. [...] We have three rooms, a kitchen and a bathroom and a small garden where I can dry the laundry and grow some flowers. [...] The heating system here is poorer than back home [but] winters are [also] much milder here and snow does not stay on the ground for long—but I still long for our furnaces that kept two or three rooms warm. I also long for snowy meadows and starry winter nights. (19 November 1956)

When Aino is decorating her apartment in Harlow, she eagerly sends Helga detailed descriptions of the furniture, wallpapers, and carpets, even including samples of wallpapers in her letter. She writes about her work environment and duties, her colleagues and friends, and her small holiday trips to Europe. Aino's relatively well-established life is disrupted when, in 1958, upon the initiative of her husband, the family moves to the United States. The move is further complicated by the worsening of Aino's health, culminating in surgery from which she had no time to recover properly before her departure to America. For about two months the correspondence between the sisters ceases, and some of Helga's letters are returned with 'Return to Sender' written on the envelope. Knowing about her sister's state of health and surgery, Helga is worried sick about Aino. 'I was quite desperate with fear', Helga writes on 16 September 1958. 'I feared that the worst had happened to you after the difficult surgery. [...] I was like a sleepwalker with eyes brimming with tears'.

When the correspondence resumes, the mode of Aino's letters has changed. Although she writes of the powerful impression of arriving in New York on the ocean liner and cruising past the Statue of Liberty, settling in yet another new country is difficult for her. Having left a well-paying job, friends, and a cosy home, in the United States she feels lonely and rootless, and at first has difficulties finding any job at all. Even though her employment situation gradually improves, her health deteriorates, as does her relationship with her husband. The feeling of homesickness that gradually started disappearing from her letters from the United Kingdom now reappears with new force; more than ever, she longs to be with her sister. In her letters, however, despite an emphasis on the lasting importance of her homeland, Aino's sense of identity starts to change towards a more transnational one: 'Over time, our lives have taken different paths. You are like an oak tree with roots deep in the native soil, I am merely a loose rock' (2 February 1958).

Helga's overviews of her life are of quite a different nature: they contain few if any details of the living conditions of her family, the jobs she and her husband have, or any changes in their life brought about by different regime in Estonia. Even from my childhood I remember quite open discussions about the nature of the regime and my family's firm resistance to it, yet in Helga's letters to her sister there is no mention of her family's disposition towards the regime, the difficulties she and her husband had finding work after the war, or the tight living conditions in an apartment that they first had to share with two other families. Reflecting on the complexities of conducting oral history interviews in post-Soviet Russia in the mid-1990s, Bertaux, Rotkirch, and Thompson argue that

> in Soviet Russia for seventy years, remembering was dangerous, not only to yourself, but to your family and friends. The less that people knew about you and your family story the better, because most information was potentially dangerous and could be twisted into material for a denunciation. (7)

Without doubt, Helga's reticence with regard to the details of even the everyday life of her family can be attributed to precautionary measures she employs to avoid causing harm to her family or circle of friends and to ensure the continuation of the correspondence with her sister. Looked at from this perspective, her words in one of her first letters—'What can I write in just one letter? I should write a book, not a letter' (1 October 1956)—acquire a different meaning. Indeed, a letter the length of a book would be needed to fill Aino in on all the changes in her family's life, yet it would be impossible for Helga to write such a 'book', and so her use of this wording may hint at the political obstacles impeding open communication.

The overview of her life that she provides in the two first letters (one sent from Finland and the other from Estonia) offers a good example. In her letter of 21 August 1956 she writes, 'I stayed home at first but I work already for four years, currently at two jobs, one of which is within my field. Eedi also works, and we are both quite busy. However, for catering for the needs of a household with three children, this is inevitable'. Given that in 1956 Helga's youngest child was already in grade five, staying at home for so long for family reasons would have been unnecessary for her, especially since she mentions the need to have two jobs just in order to get by.

Helga's letter hints at the difficulty in finding work in her field of competence (law). In reality, the main reason it was difficult for her to find employment after the war was the nature of her profession. Law was considered a political field by the Soviet regime, and specialists trained by the previous regime were considered unsuitable for employment within the Soviet system. There is no mention in the letter of what kind of jobs Helga and her husband may have had or to living conditions or their home, other than the inclusion of their address. Helga's second letter, sent from Estonia, contains even fewer references to her life, apart from the rather general comment, 'Like you, we have had a harsh confrontation with life during the postwar years [...] but so far we have been able to gain an upper hand' (1 October 1956). The rest of the letter focuses on the progress of her children at school, a topic that Helga feels safe in elaborating on in detail, both here and in her future letters.

Never Fully Mine

In *What They Saved: Pieces of a Jewish Past,* Nancy K. Miller, reading the courtship letters of her parents from the summer of 1934, writes,

> This story of that young couple in love feels so close. And yet no matter how close I can get to the spot, stop, linger there—the people they were that hot summer forever escapes me. But their letters have become mine, and reading them, I'm melting too. (93)

When I started reading the letters of my grandmother and great-aunt, I rejoiced at the extensive textual corpus that for me as a textual scholar is a way of accessing the world that I am well accustomed to and comfortable with, and organising the letters gave me great emotional and intellectual satisfaction. The correspondence can be viewed as a chronicle of my family's life, across the span of nearly half a century, containing valuable information that no one else in my family would remember in detail. I also assumed that reading through all the letters would provide me with full access to the lives of the sisters as my family members and foremothers. My grandmother Helga, or Memm as we called her, was my main caregiver throughout my childhood and youth, and she seems absolutely, taken-for-grantedly familiar and close. My great-aunt Aino, as Memm writes in one of her letters, 'was a topic of conversation in our family on a daily basis, a real family member' (12 June 1962), so it felt as if I had always known her well, too. Yet the correspondence, where they are so intensely tuned toward each other, also forms an unreachable self-contained unity, testifying to the utter dedication of the sisters to each other, and their creation of a private world of words, images, memories, and dreams to which only they could have full access.

For me, the process of reading their letters has been, in the words of Maarit Leskelä-Kärki, a lesson in 'dealing with otherness and accepting that we can never fully reach people through this process and reveal every part of their lives' (31). The letters may offer me a glimpse of the people Helga and Aino were, from the late 1950s through the late 1980s, but their letters can never become fully mine, as the most important aspect of the correspondence is the intimate dynamics of the relationship itself. The letters are a record of a lifetime of painful longing—of settling for words on paper instead of a real conversation, of going over a photograph with a magnifying glass instead of having a real-life encounter—but they also provide powerful evidence of individual people's determination to build and maintain intimate family relationships in spite of separation caused by the forces of history.

Funding

This work was supported by Estonian Science Foundation grant ETF9035, 'Dynamics of Address in Estonian Life Writing', and IUT22-2, 'Formal and Informal Networks of Literature, Based on Sources of Cultural History'.

Notes

[1] A few examples of life story collection in Estonia are the three-volume *Eesti rahva elulood* [Life Stories of Estonian People], which contains life narratives of Estonians of different generations with different life paths and which was published from 2000–2003 (Hinrikus, ed.); a collection focusing on deportation and *Gulag narratives, Me tulime tagasi* [We Came Back] in 1997 (Hinrikus, ed.) and a collection of life stories in exile, *Rändlindude pesad: Eestlaste elulood võõrsil* [The Nests of Migratory Birds: Life Stories of Estonians Abroad] in 2006 (Kirss, ed.). Collections of Estonian life stories in English include *She Who Remembers Survives: Interpreting Estonian Women's Post-Soviet Life Stories* (Kirss et al. 2004), *Estonian Life Stories* (Hinrikus and Kirss 2009), *Carrying Linda's Stones* (Lie et al. 2006), and *Soldiers of Memory: World War II and Its Aftermath in Estonian Post-Soviet Life Stories* (Kõresaar 2010).

[2] In addition, there are also a couple of letters from the late 1940s sent by Helga and their mother, Emilie Valgerist, to Scotland, where Aino lived and worked right after arriving in the United Kingdom. There are no letters from Aino from Scotland, although Aino confirmed in a letter to a relative in the United States that she received the letters and also responded to them. As Helga preserved Aino's letters with ultimate care, it is unlikely that the letters from the 1940s would have been lost but rather that Soviet censorship officials confiscated them.

[3] The mass emigration from Estonia from August to October 1944, also referred to as the 'Great Escape to the West', was brought about by the offensive of the Soviet troops in World War II (Kumer-Haukanõmm 'Eestlaste Teisest' 16). As the memory of the first Soviet occupation (1940–1941), which brought with it a wave of arrests, murders, and the deportation of more than 10,000 people, was still fresh, people fled in fear of further repressions. The overall number of people who fled to the West during World War II is estimated between 70,000 and 90,000, including children and those who perished during the journey (Kumer-Haukanõmm 'Eestlaste põgenemine' 17). Aino emigrated because her husband was German by nationality and had been drafted into the German Army during the German occupation. Aino married Paul, her high school sweetheart, right after graduation at the age of 18 before he left for military service. She was 21 years old when she left Estonia. She emigrated with the hope of being able to reunite with her husband but was unable to obtain any information about him for many years. Aino's husband was a Soviet POW, and the couple finally met in the late 1940s in the United Kingdom.

[4] In his study of everyday life in Stalinist Estonia, Olaf Mertelsman highlights the importance of kinship ties in Estonian society, which remained relevant during the Soviet period as well. Mertelsman ranks the hierarchy of relationships in the following manner: '[T]he most important was a family member, then a friend, a schoolmate or comrade from military service, a relative, an acquaintance, a person from the same village, and finally, a person from the same region' (103).

References

Archival Source
The correspondence of Helga Sitska and Aino Pargas, 1947–1989. Private collection.

Barton, David, and Nigel Hall. 'Introduction'. *Letter Writing as a Social Practice*. Ed. David Barton and Nigel Hall. Amsterdam: John Benjamins, 1999. 1–14.

Bertaux, Daniel, Anna Rotkirch, and Paul Thompson. 'Introduction'. *On Living through Soviet Russia*. Ed. Daniel Bertaux, Anna Rotkirch, and Paul Thompson. New York: Routledge, 2004. 1–23.

Cockin, Katherine. 'Ellen Terry, the Ghost-Writer, and the Laughing Statue: The Victorian Actress, Letters, and Life Writing'. *Journal of European Studies* 32.2–3 (2002): 151–63.

Cook, Elizabeth Heckendorn. *Epistolary Bodies: Gender and Genre in the Eighteenth-Century Republic of Letters*. Stanford, CA: Stanford UP, 1996.

Cvetkovich, Ann. *An Archive of Feelings: Trauma, Sexuality, and Lesbian Public Culture*. Durham, NC: Duke UP, 2003.

DeHaan, Kathleen A. 'Negotiating the Transnational Moment: Immigrant Letters as Performance of a Diasporic Identity'. *National Identities* 12.2 (2010): 107–31.

Elliott, Bruce S., David A. Gerber, and Suzanne M. Sinke. 'Introduction'. *Letters across Borders: The Epistolary Practices of International Migrants*. Ed. Bruce S. Elliot, David A. Gerber, and Suzanne M. Sinke. Houndmills: Palgrave, 2006. 1–25.

Gerber, David A. 'Epistolary Masquerades: Acts of Deceiving and Withholding in Immigrant Letters'. *Letters across Borders: The Epistolary Practices of International Migrants*. Ed. Bruce S. Elliot, David A. Gerber, and Suzanne M. Sinke. Houndmills: Palgrave, 2006. 141–157.

Hanna, Martha. 'A Republic of Letters: The Epistolary Tradition in France during World War I'. *American Historical Review* 108.5 (2003): 1338–61.

Helbich, Wolfgang, and Walter D. Kamphoefner. 'How Representative Are Emigrant Letters? An Exploration of the German Case'. *Letters across Borders: The Epistolary Practices of International Migrants*. Ed. Bruce S. Elliot, David A. Gerber, and Suzanne M. Sinke. Houndmills: Palgrave, 2006. 29–55.

Hinrikus, Rutt, ed. *Me Tulime Tagasi*. Tartu: Eesti Kirjandusmuuseum, 1999. Print.

Holland, Patricia. 'History, Memory, and the Family Album'. *Family Snaps: The Meanings of Domestic Photography*. Ed. Patricia Holland and Jo Spence. London: Virago, 1991. 1–14.

Jolly, Margaretta. *In Love and Struggle: Letters in Contemporary Feminism*. New York: Columbia UP. 2008.

Jolly, Margaretta, and Liz Stanley. 'Letters as / not a Genre'. *Life Writing* 2.2 (2005): 91–118.

Kirss, Tiina. 'Põgenemisteekonnad ja põgenemislood'. *Rändlindude pesad: Eestlaste elulood võõrsil*. Ed. Tiina Kirss. Tartu: Eesti Kirjandusmuuseum, 2006. 611–46.

Kirss, Tiina, Ene Kõresaar, and Marju Lauristin, eds. *She Who Remembers Survives: Interpreting Estonian Post-Soviet Life Stories*. Tartu: Tartu University Press, 2004. Print.

Kirss, Tiina, and Rutt Hinrikus, eds. *Estonian Life Stories*. Budapest: Central European University Press, 2009. Print.

Kõresaar, Ene, ed. *Soldiers of Memory: World War II and its Aftermath in Estonian Post-Soviet Life Stories*. Amsterdam, New York: Rodopi, 2011. Print.

Kumer-Haukanõmm, Kaja. 'Eestlaste põgenemine Saksamaale.' *Eestlaste põgenemine Läände Teise maailmasõja ajal*. Ed. Terje Hallik, Kristi Kukk, and Janet Laidla. Tartu: Koproratsioon Filiae Patriae, 2009. 13–54.

Kumer-Haukanõmm, Kaja. 'Eestlaste Teisest maailmasõjast tingitud põgenemine läände'. *Suur põgenemine 1944: Eestlaste lahkumine Läände ja selle mõjud*. Ed. Kaja Kumer-Haukanõmm, Tiit Rosenberg, and Tiit Tammaru. Tartu: Tartu University Press, 2006. 3–38.

Lie, Suzanne Stiver, Lynda Malik, Ilvi Jõe-Cannon, and Rutt Hinrikus, eds. *Carrying Linda's Stones: An Anthology of Estonian Women's Life Stories*. Tallinn: Tallinn University Press, 2006. Print.

Leskelä-Kärki, Maarit. 'Constructing Sisterly Relations in Epistolary Practices: The Writing Krohn Sisters (1890–1950)'. *Nordic Journal of Women's Studies* 15.1 (2007): 21–34.

Liiv, Juhan. *The Mind Would Bear No Better: A Selection of Poetry in Estonian and English*. Trans. Jüri Talvet and H. L. Hix. Tartu: Tartu UP, 2007.

Maybin, Janet. 'Death Row Penfriends: Some Effects of Letter Writing on Identity and Relationships'. *Letter Writing as a Social Practice*. Ed. David Barton and Nigel Hall. Amsterdam: John Benjamins, 1999. 151–78.

Mertelsman, Olaf. *Everyday Life in Stalinist Estonia*. Frankfurt am Main: Peter Lang, 2012.

Miller, Nancy. *What They Saved: Pieces of Jewish Past*. Lincoln: U of Nebraska P, 2011.

Morrison, Toni. 'The Site of Memory'. *Inventing the Truth: The Art and Craft of Memoir*. Ed. William Zinsser. New York: Houghton Mifflin, 1995. 83–102.

Stanley, Liz. 'The Epistolarium: On Theorizing Letters and Correspondences'. *Auto/Biography* 12 (2004): 201–35.

The Odyssey Quilts: Narrative Artworks of Childhood, War, and Migration

Nonja Peters

This chapter focuses on a set of three wall hangings, known as the Odyssey Quilts, that are held in the Power-House Museum in Sydney, Australia.[1] They present creative visual narratives—made up of appliqued quilting pieces—portraying their creators' recollections of their experiences of childhood, wartime, and migration. The quilts are the work of ten Dutch women who immigrated to Australia after the Second World War: Gerada Baremans, Johanna Binkhorst, Yvonne Chapman, Ann Diecker, Anna Dijkman-Tetteroo, Ineke McIntosh-Eichholtz, Frances Larder, Vera Rado, Francis Widitz, and Vicky van der Ley. Five of them came from the Netherlands (NL) and five from the Netherlands East Indies (NEI), now Indonesia.[2] Drawing upon their family histories, as well as their personal memories and mementos, the women produced artists' biographies as a resource for their collective endeavour. These written narratives, along with photographs and other records that were aggregated into visual diaries, provide personal insights and stories that serve as a valuable contextual framework for interpreting the visual images embedded in the quilts. While the quilting pieces draw upon memories that are individual and personal, they communicate a larger story of post–Second World War migration to Australia, releasing, in condensed visual form, histories and contexts that the artists' biographies present more expansively in words. In this chapter the visual diaries and supporting autobiographical material are used as an aid to the interpretation of the quilts as personal expressions whose relevance extends far beyond the individual stories they tell (Figures 1–3).

Overall, the quilting project took five years to complete.[3] The principal aim driving the quilters' undertaking was to leave a lasting legacy, through visual impressions of these defining life experiences, for their grandchildren and for future generations. Their vision was realised in 2008 when the quilts were acquisitioned into the permanent collection of the Power-House Museum in Sydney.

Each of the three quilts—referred to here as the childhood quilt, the wartime quilt, and the Australia quilt—has in the background a set of evocative symbols that the women chose to further contextualise their stories. These symbols function as an additional interpretive layer, as Frances Larder explains:

Figure 1. Childhood quilt: Memories of childhood in the Netherlands or Netherlands East Indies

> We chose ocean waves and flying birds for the "Childhood quilt" to preempt symbolically our future departure from our "Motherland," where, in retrospect, we had spent what seemed to us many years later an idyllic childhood. The background images on the war quilt are a cross, symbolizing all the war deaths, and the wings of a dove, the sought-after peace. The background symbols on the quilt dedicated to first impressions of Australia are the "Southern Cross" and a kangaroo.

The latter represent the wide-open spaces of Australia to reflect the expected lifestyle change to a better future for these families traumatised by war and having to flee their country of origin by taking yet another major step, including all the risks and opportunities associated with migrating to a new homeland.

Of special note is the variety of shapes the quilting pieces take. For instance, the women's lovingly recalled childhood experiences and first impressions of an awesome and strange Australia are expressed in picture format: portrait or landscape. In contrast, their wartime experiences—and here I must add that only six of the women contributed to this quilt, the others not wanting to have to reengage with this traumatic period in their lives—are conceptualised in mushroom shapes, much like the text balloons found in a cartoon or comic book. The women say they chose this shape because for them it exemplified the 'fractured fragments' that characterised their wartime recollections compared to the relative coherence of their memories of childhood and migration. This change raises the question of how much people retain or want to retain of the traumatic events that they survive. Like any other artwork or museum exhibit, the women's depictions on each quilt signify or refer to many things beyond it, including the person or collective who made it, the particular time and place in which it was created, the technology of the period, and the prevailing social situation. Moreover, these have been influenced by a sweep of historical events: local, national, and international. What we see is the tip of an iceberg, which in holistic

Figure 2. Wartime quilt: Memories of the Second World War

terms is equal to only a fraction of the object's history. Many of the characteristics that make the object what it is are hidden from our view. The Odyssey Quilts represent such an object, for although they reflect fragments of memories—positive as well as traumatic—of significant events and periods in the lives of the women who created them, they symbolise a much larger set of experiences with which many others can also identify.

To comprehend the quilters' motives and experiences, it is essential first to appreciate the critical impact of the differences—in cultural and geographical terms—between the experiences of the women in the team who spent their childhood and wartime in the Netherlands as compared to those who spent their childhood and war years in the Netherlands East Indies. While the two societies were similar in being highly stratified, the contexts in which they functioned; the bureaucratic structures, policies, and procedures that governed daily living; as well as the physical environment and climate were vastly different. For example, consider life in a tropical, socially stratified colonial environment, with colonists at the top of the status tree and with servants to attend to one's every whim, and compare it to life in a temperate climate

Figure 3. Australia quilt: Memories of arrival in Australia

controlled by a socioreligious, cradle-to-grave, pillarized (Verzuiling) infrastructure dominated by elites with social mobility. They were markedly different, and the wartime experiences of the two groups were also very different. Most notably, in wartime the occupying forces were from different countries: in the Netherlands East Indies the oppressor was the Japanese; in the Netherlands the oppressor was the Nazi regime (Peters 2010).

The quilting project was conceptualised and organised under the leadership of Frances Larder, who also recruited the nine other women willing to commit to a public sharing of some of the most pleasurable, emotional, and distressing aspects of their lives.[4] The process of creation had the women delve into their family of origin's tangible heritage—diaries, letters, film clips, and photographs—as well as intangible heritage—oral histories and family folklore, myths, and legends. The women noted the importance of these *aides memoire* in helping them recapture lost memories. From this process the women produced an artist's biography and a visual diary from which they were able to (re)imagine and (re)produce a mind picture that best expressed their childhood experiences, their first impressions of Australia, and poignant aspects of wartime, which they would ultimately transform into their quilting pictures.[5]

Underlying my analysis of the women's quilting pictures and visual diaries are two broad questions. First, to what extent do these visual narratives express, reflect, and transmit the intangible cultural heritage that shaped the women's own and their families' histories, given the historical context, especially the Second World War, and that most of them were children when war broke out? Second, how does this heritage contribute to their sense of place, identity, and belonging in their host environment, and to what extent do their quilting pieces demonstrate that it is possible to develop emotional bonds to a new homeland, its environment, and its cultural heritage?

The *Macquarie Dictionary* defines *cultural heritage* as 'that which comes or belongs to one by reason of birth; an inherited lot or portion; or something reserved for one' (831). We often refer to material possessions in discussions about cultural heritage and in relation to community historic buildings; archaeological sites; and artefacts held in museums, archives, and libraries. Yet cultural heritage can have much broader connotations and significance. The Odyssey Quilting Project is about mapping memory and transposing this 'intangible cultural heritage'—the women's recollections of central life experiences—into visual imagery. This process relies heavily on what the women remembered and the nature of the legacy they wished to leave. For philosopher James Booth, memory (accepting its limitations) 'is centred on an absence, tries to make it present, and in doing so answers the call of the trace' (114). The idealisation of childhood, in this context, is also possibly linked to the process of 'reflection' that inevitably had the quilters comparing their childhoods to their later experiences of war and migration. A consequence of this process of reflection and comparison is that their childhood is expressed visually as that uncomplicated period in their biography when life is—literally—child's play.

Childhood Quilt: Memories of Childhood in the Netherlands or Netherlands East Indies

In this section I focus on the visual diaries and quilting pieces of Frances Larder (Figure 4), Ineke McIntosh-Eichholtz (Figures 5 and 7), Gerada Baremans (Figures 6 and 8), Vicky van der Ley (Figure 11), and Johanna Binkhorst (Figures 9 and 10).[6] The following extract from Ineke's artist statement illustrates how positively the women reimagine this 'carefree' period and some of the poignant relationships with the Indigenous people who featured in it at the time, as well as love of the environment:

> The first ten years of my life were spent in the Dutch East Indies living in Batavia (Djakarta) and Buitenzorg (Bogor). […] [M]y early memories mainly go back to my primary school years. To escape Batavia's oppressive heat, the school holidays would be spent in little mountain villages. The names *Tjipajung* and *Tjimbuluwit* immediately come to mind. Here, time spent in the rivers with their big boulders and swimming holes became the highlight of our holidays. Those carefree years are still accompanied by memories of a gentle people, the unforgettable landscape and its exotic plants, as well as the ever-present air of spicy aromas.

Ineke's visual diary of her childhood in the Netherlands East Indies contains only three images that reflect Dutchness: the family cat sitting in a space that can be described as a Western family home, a letter signifying contact with relatives from the mother/fatherland (Holland), and a Poëzie booklet (poetry). The Poëzie booklet is a common object in a middle-class Dutch girl's life; it's where family and friends attribute a poem they

Figure 4. Rice fields in Java on the way to school (Frances)

believe to signify her character. Ineke's visual diary also has two religious icons. The first is the Wayang in the far-right corner, a symbol associated with shadow puppetry.[7] The focus of Wayang theatre is mainly Hindu religious epics, depicting the Mahabharata and the Ramayana. The other is a bust of the Buddha, such as those found at the famous religious site of Borobodur near Surabaya, East Java.[8] The paper fan appears to depict Chinese or Japanese calligraphy. These images are surrounded by images of the natural environment, the 'place' described by Ineke in her artist statement: Javanese trees; leaves; flowers; and mountains with a river running through them swollen by recent rain carrying all manner of plants to the flatlands, rocky pools, paddy fields, and palm trees. Frances also recalls the natural environment that was familiar in her youth in her depiction of the rice fields that she passed on her daily walk to school in the Netherlands East Indies (Figure 4).

Not all the imagery in Ineke's and Gerada's visual diaries find a place in their quilting pieces. However, the visual diaries make a significant contribution in that they make clear statements about the differences between the societies that permeate their imagery. Whereas Gerada's diary shows a controlled landscape, where the physical and built environment is dominated by man-made structures and patterns, Ineke's Javascape has meandering streams and swaying coconut palms. Gerada's visual diary depicts the water fountain that supplies the town's water and a local form of transport,

Figure 5. Ineke's visual diary

Figure 6. Gerada's visual diary

Figure 7. Ineke's quilting piece

Figure 8. Gerada's quilting piece

via man-made canals. The family home is a solid brick structure. Tiny Brownie camera photos show her siblings dressed in stiff, starched Sunday-best clothing. The rural provincial landscapes that figure prominently in these diaries are vastly different from each other. Ineke has chosen 'ferrying flowers from the hothouses to the market' as the subject of her quilting piece to represent how she perceived her childhood in the Netherlands.

In Ineke's quilting piece she highlights the peaceful exotic Java scene of her youth through the juxtaposition of two contrasting modes of transport: the iconic Becak, which functions mainly for internal travel and tourism, and a migration symbol—an ocean liner comparable to the ones that would eventually transport her forever out of her country of birth to the Netherlands and after that Australia. Ineke's homeland—her country of birth—was lost to her when the Netherlands East Indies became the Republic of Indonesia in December 1949. Gerada, on the other hand, did not have to grapple with the same loss of place because when Nazi Germany retreated from its wartime occupation of the Netherlands, the country was free and Dutch again.

Johanna Binkhorst's 'Monday washday' conveys a sense of the order and ordinariness of everyday life in pre-war Holland—peaceful, serene, and homey—holding the underlying surety that the north wind would dry the wash. While you see only the clean wash go online, the imagination recalls the labour-intensive commitment needed to achieve

Figure 9. *Johanna's quilting pieces*

this penultimate action of hanging it on the line before removing it for ironing. Like a museum object that has most of its history hidden, doing the washing in the 1930s required chopping wood to light the copper; soaking; washing and scrubbing; rinsing in Reckitt's Blue; and starching shirts, collars, and dresses. The process took all day to complete, yet these activities are hidden from view. Her other quilting piece depicts village life, with the family home and church prominently situated among beautiful trees and gardens and the ubiquitous barge proudly brandishing the Dutch flag as it sails down the river or canal. The childhood pieces of all the quilters portray the natural environment as well as buildings and structures that were important aspects of their daily lives.

Barbara Bender argues that we are only capable of understanding the world around us, at least initially, from what we have learned, been exposed to, and received in the way of narratives, traditions, and beliefs (4). On the other hand, experiencing 'place' through the body in the way Ineke implies is central to de Certeau's philosophy that the body, in movement, gesticulation, walking, and taking its pleasure, is what indefinitely organises a 'here' in relation to an abroad, a 'familiarity' in relation to a 'foreignness' (de Certeau, 130).

Gupta and Ferguson suggest that

> [t]he ability of people to confound the established spatial order, either through physical movement or through their own conceptual and political acts of re-imagination,

Figure 10. Johanna's quilting pieces

means space and place can never be "given" and that the process of their socio-political construction must be considered. (17)

This prompts the question for Gupta and Ferguson: Why do we become attached to a place?

Kevin Lynch claimed it is the process of creating the man-made environment—nodes, paths, edges, and districts—that marks out a sense of place, creating an understanding of one's environment that, at least in navigational terms, engenders a 'sense of emotional security' (Lynch 7). Following Lynch, Norberg-Schulz makes the point that 'place' is defined more by its ability to serve as a 'habitat' for its residents than by its physical properties. This has led him to describe the connection between humans and their homeland as more spiritual than physical, relying on senses, memories, and beliefs (5). For Norberg-Schulz, experiencing a place deeply enables people to bond with it, to develop connections, emotional attachments, and meanings that are relevant with regard to developing their sense of belonging and identity. Norberg-Schulz's and de Certeau's philosophies relating to the spiritual and emotional dimensions of place are reflected in the women's quilting pieces about their childhood bonding with their country of origin and the bonds they subsequently forge with Australia.

Figure 11. Fishing with father (Vicky)

The bonding process described by Norberg-Schulz is evident in the quilting piece by Vicky van der Ley, who was born in the small village of Sliedrecht, near Dordrecht in the south of Holland. She recalls the place of her fondest memory of childhood, 'the Merwede River that ran near our house. My brothers and sisters and I would swim, row our boat, or skate on its frozen surface in winter'.

De Certeau's and Bender's philosophies of how we develop the bonds that constitute our 'sense of place' are articulated visually in the diaries and quilting pieces of Ineke and Gerada and the quilting pieces of Frances Larder, Johanna Binkhorst, and Vicky van der Ley. The manner in which these theorists construct the experience of developing a 'sense of place' raises the prospect that it is possible to bond with multiple places. I am thinking specifically here of immigration in terms of emigrants' capacity to integrate with the receiving society and its environment to create a new 'sense of place'. The serene childhood worlds portrayed by Frances, Jo, Ineke, Gerada, and Vicky were shattered by war, invasion, occupation, revolution, repatriation, and migration. Even so, their visual diaries and quilting pieces about Australia show a bonding-type process unfolding that could be deemed an unconscious determination to lessen the traumatic impact of loss and diaspora of their previous lives.

Figure 12. Graveyards, Second World War (Johanna)

Wartime Quilt: Memories of the Second World War

The wartime quilt was in fact the third quilt the women produced, but for the sake of historical chronology I present its analysis second. The wartime quilt assumed the shape of a kimono by chance. Since only six of the ten women were prepared to work on it, they began the process of producing it in three panels and then noticed how this had taken the shape of a kimono. However, it also contains many quilting pieces by women whose experience of war was in the Nazi-occupied Netherlands. The main themes of the artists' statements, visual diaries, and quilting pieces are focussed on invasion and occupation; living under a foreign oppressor; incarceration in internment camps; the search for adequate food, fuel, and clothes; and the fear, brutality, violence, disease, and death that characterised their everyday life at that time. A notable distinction between the Dutch in the Netherlands compared to those in the Netherlands East Indies was that the majority of the Dutch population who endured the war under Nazi occupation from 1940 to 1945 remained living in their own homes throughout that time (Peters *From Tyranny*) (Figures 12 and 13).[9]

Johanna Binkhorst's quilting piece (Figure 16) recalls the Nazi incursion into the Netherlands. She writes,

Figure 13. Bowing to the Japanese, Second World War (Frances)

On 10 May 1940, Germany declared war on the Netherlands. I had been married for three days and what should have been beautiful and happy times ahead became a time of sorrow and frustration. My memories of five years of Occupation are of blocking off our windows every night with black paper before curfew began at 8pm, rations of essentials, boredom, and not enough food, which became worse over the war years—especially fuel for keeping warm.[10]

The German forces had invaded the Netherlands despite its policy of neutrality and without a formal declaration of war. Despite continuing to live in their own homes, the lives of the Netherlands' Dutch changed in every other way. People lived in fear, for they were no longer free. Some families also had to accept Nazi military personnel being billeted into their homes.[11]

The Nazis had four goals for the Netherlands: transforming it into a national socialist state; exploiting the economic potential of Dutch industries and the labour force;

Figure 14. Japanese invasion of the Netherland East Indies, March 1942 (Frances Larder). See bottom of the middle panel of the wartime quilt (Figure 2); it shows this quilting piece in its entirety

purging the Netherlands of all Jews; and preventing all aid to Germany's enemies through espionage, sabotage, and guerrilla activity (Swarthmore.edu). From 1943, when the Nazis had introduced the obligation to work, every male in the Nazi-occupied zones between 18 and 50 years of age and every unmarried woman between 21 (later 18) and 35 years of age could be conscripted to work for the German Reich (Das Bundesarchiv). However, men in Dutch Military Forces when the Netherlands capitulated were now designated prisoners of war (Figures 14 and 15).[12]

Frances Larder's quilting piece (Figure 14) conveys the sense of power that accompanied the Japanese invasion of the Netherlands East Indies. Vera Rado's description of the arrival of the occupying forces into Batavia makes it palpable:

> It was a black day, 8 March 1942, in more than one sense. The oil tanks on the southwestern edge of Batavia (present-day Djakarta) the city were blown up by the Dutch to prevent the precious fuel from falling into the hands of the enemy. From early morning there was a huge pall of black smoke hanging over the city, and against this ominous backdrop we watched the occupying army's progress through our streets. First the tanks with their red and white flags then armoured carrier, trucks, then masses of soldiers on foot, and on bicycles. They looked triumphant, but we were trembling with apprehension peeking through the louvres of our front door and windows. What

Figure 15. Nazi invasion of the Netherlands, 10 May 1940 (Johanna)

would happen to us? We were totally at their mercy—no laws, no constitution, no army or police to protect us. ('Life Story')

From the beginning, intense Japanisation of the population was enforced, especially in schools, where pupils were duty-bound to be loyal to Japanese symbols and ideology. Currency and the annual calendar years also changed to Japanese. All overtly political organisations were dissolved, and the most prominent pre-war nationalist leaders were immediately released from captivity to be incorporated into the administrative structure. In these positions, nationalists were used as propaganda tools to spread the gospel of the Asia Co-Prosperity Sphere and Greater East Asia slogans of 'Asia for the Asians' and were directed to carry out various Japanese projects (De Bary 622).

In contrast to the Nazis, the first undertaking by the Japanese Occupation Forces in the Netherlands East Indies was to categorise the Dutch as Caucasians or Eurasian. Caucasians were rounded up and incarcerated for the duration (three and a half years) of the Pacific War, 1942 to 1945. To comply, Dutch people immediately had to register at the town hall and obtain an identity card, which they had to carry on them at all times. Cars were confiscated, radios had to be handed in to be sealed so that only local stations could be received, and very soon all public servants from

Figure 16. Wartime image (Wilhelmina)

the governor general down to the most junior clerk were rounded up and imprisoned. The imprisoned also included all male teachers, so school ceased altogether (Krancher).

Buitenkampers

Although the majority of *Indisch* Dutch (Eurasian persons of Dutch Indonesian or Dutch Chinese descent) lived outside internment camps (hence the label *buitenkampers*, literally meaning 'living outside camps'), they too were treated appallingly by the Japanese, who, upon noting their generally stronger allegiance to their Dutch rather than Indonesian heritage, harassed them remorselessly, often brutally, on this basis (Helmrich). This included denying them access to employment and their savings. Bank accounts were frozen, homes were requisitioned, and most lost their jobs as soon as an Indonesian had been trained to handle their position. To survive, they sold all they had, and many women resorted to prostitution. Frances and her family were among the few Caucasians left outside the camps, as her mother, Wilhelmina de Brey, explains:

Figure 17. Wartime image (Wilhelmina)

> I was fortunate to not be interned. My son had a heart condition, and my doctor told me that if my children and I went into a camp he would die. I decided to stay out of the camps at any cost…. It was soon apparent that we were all penniless, so we decided to start a craft group and I would go around and visit other Dutch people who were still working under the Japanese. But soon those people disappeared into the camps, too. We had to rely on the help of Dutch-Indonesian (Eurasian) people, and I received all the help that was possible. This was dangerous, because if caught supplying food, the soldiers issued summary justice. The hate and hopelessness that we felt during the invasion of our home by the Japanese officers is hard to imagine. Time has softened those feelings, but they will never be forgotten.

When the Japanese entered Batavia, Wilhelmina's husband (Frances's father) was taken prisoner. As a consequence, she was left to cope alone, pregnant, with three children, living under permanent Japanese house arrest without access to basic necessities. She overcame this situation by breaking the seal at night to obtain the items needed to sustain them. The Japanese also allowed the family's Indonesian nursemaid and friend to stay with them, and she would sometimes get them food from the *kampong*.

Wilhelmina De Brey notes of her contribution to the war quilt (Figures 16–18),

Figure 18. Wartime image (Wilhelmina)

> I created three images of this [war]time. The pieces are 'in memory' of those I 'failed to save' through the betrayal of … a Dutch woman [who] came to visit our house and turned out to be a very close friend of a Japanese officer, and from this we had a premonition of danger ahead. [Wilhelmina had joined the underground sometime earlier, and it seems that the woman visitor informed the Military Police, the Kempeitai, about the group's underground activities.] Shortly after her visit to their house, the Kempeitai invaded. Only my youngest friend was allowed to stay in the house with her little girl and my children. When I said good-bye to my children, my eldest son was so upset he got a high fever from fright. I asked my friend to look after him. Not knowing that one of the Japanese officers understood Dutch, he asked my friend how many of the children were hers and she answered that she had only one but that I had four. Thankfully, the officer decided that I should stay in my house.

Wilhelmina's war quilting contains images of the leaf of the castor oil plant to recall the seeds of this plant that the locals had to supply to the Japanese for them to press to extract the oil.[13] In recalling that period in her life, she writes, 'My home with its blacked-out windows. The Japanese planes would fly over at night, and it was very frightening. Later, the Americans would drop supplies of peaches, chocolate, and corned beef.' Also depicted in the quilt are the wheels of her bike, representing how she would travel to the different houses to sell her crafts to buy food and distribute news from forbidden radios. She explains how she did this (Figure 19):

> I would deliver the news on my bike to all our friends. The news was written on [rice] paper, which was folded up and hidden inside a beautiful Chinese ring, which I still have. It has a large red stone, and the top opens, hiding a small compartment.

The invasion of Wilhelmina's home happened only six months before the end of Japanese rule. The fact that there were people whose lives she could no longer protect gave her great grief, especially since, as she points out, 'The war was almost behind us'.

Caucasian Dutch

In contrast Vera Rado and her family were interned:

> My brother and I had to go to Council chambers in the morning to fill out forms, and when we returned home at lunchtime, our father had already been taken to prison by Japanese soldiers. We were ordered to pack and be ready to be interned also. The delay had afforded my mother the time to figure out what to take in the smallish suitcase we were each allowed. She had the presence of mind—for which I praise her to this day—to pull out the bottom drawer of her dressing table and upend it into her suitcase. It contained patent medicines, including quinine and sulfa tablets. Her act of foresight saved not only my life but that of a few others. (Rado *In Japanese Captivity*, 1–11 [extracts]).

Vera and her family were at first locked up in the Werfstraat Jail. I quote from her 'Life Story':

The compound to which we were taken was surrounded by high stone walls topped with broken glass. There were six large cells with barred doors and huge copper padlocks. Each cell was meant for ten to twelve prisoners, but we were herded into

Figure 19. The ring Wilhelmina used to hide news

Figure 20. Locks symbolise the locked gates of Adek camp (Vera)

> them with about thirty-five to forty women and children. There was a hole in one corner for a squat-down toilet and mats on the stone floor for us to sleep on. At six p.m. the doors were banged shut, and with a sharp click of the padlocks, we were left in no doubt as to our status. We were prisoners of the Japanese. For how long? We barely slept that night on the cold hard floor. The noise of children crying and mothers shouting and wailing was like something out of a nightmare. There was no privacy at all, and the single toilet soon became a source of continuing stench….

From the Werfstraat jail the men were moved to Tjimahi (now Cimahi), West Java. The women and children were taken to Tangerang, twenty kilometres west of Jakarta. In February 1945 the women and children were again moved, this time to Camp Adek in Jakarta, where they stayed until the end of the war.[15] Camp Adek was part of Tjideng, the most infamous internment camp in Jakarta. Its supreme commander was the "lunatic" captain Kenichi Sonei, who terrorised the inmates by beating and hitting women when the moon was full. He was executed in 1946 for the war crimes he committed against internees (Figures 20 and 21).[16]

Survival themes and strategies dominate the quilters' biographical texts and quilting pieces relating to wartime. Fear of beatings, brutality, violence, disease, devastation, death, bombing, and lack of food and medicines—and for those in the Netherlands, fear of the biting cold without fuel to warm their homes—are prominent in the wartime memories.

War sounds had a lasting impact on Gerada, as she explains (Figure 22):

> The sound of sirens made me so frightened that, even now, I can hear the sound. Pieces of shrapnel were flying everywhere when the bomb came down. I always went under the table or hid behind the toilet. My father had a radio behind his chair and was listening to the English broadcasts, but if the Germans found this out you were shot. Can you imagine how my mother felt? One evening I found the radio and joyfully told my father. He took an axe and smashed it to pieces.

I, too, can attest to the lasting impact of such experiences. In Australia, decades after the war had finished, I watched my mother dive under a coffee table when our house was struck by lightning. She had also experienced Nazi occupation.

Johanna Binkhorst offers another perspective on war when she highlights the 'ingenuity and imagination that it generated': '[N]othing was wasted…. Pushbike tyres were made of rubber garden hose and worn shoes would be repaired with wooden soles.' Clothing shortages posed an additional challenge. Gerada's resourceful mother made clothes for her children out of the parachute silk that their father brought home from

Figure 21. Bowing to the Japanese (Vera)

his night-time forages. Anna Dijkman-Tetteroo's father cut off the top front of her brother's shoes to accommodate his growing feet. This event is central to one of her quilting pieces (see Figure 23). Anna had also to make do first with very old ladies' boots and later with wooden soles with straps on the top. These 'shoes' were particularly uncomfortable when Anna assisted during her family's fuel crisis by going to the Scheveningen Forest

Figure 22. Nazi soldiers patrolling the curfew at night (Gerada)

(Figure 24) with a girlfriend twice a week to cut wood. She notes, 'We were only thirteen years old but we managed to get thin trees down. We had a small cart to transport the wood'. Anna describes an unexpected encounter she and her friend witnessed in the woods:

> At the edge of the forest were the launching ramps of the V2s, the Flying Bombs, which were fired by the Germans to bombard Britain and later the harbour city of Antwerp. It was very scary when we could not hear the noise fading away in the distance. We then knew that the shot had failed and the V2 was coming back to the place where

Figure 23. Her brother's shoes (Anna)

it was let off. We had to dive for cover. But we both found it a big adventure, and we continued to gather wood until the end of the war.

Searching for ways to get enough food, shoes, and textiles dominate the women's wartime quilt pieces. Gerada recalls,

> My mum had to beg for food. It was very degrading, but you need to stay alive and there was no other means of getting it. Some people wouldn't give it to her, only if she took us kids with her…. My father often went out at night to get coal, wood, old shoes, anything that would burn to keep us warm. We were so scared that he would be picked up by the Germans; no one was allowed out after dark.

The lack of food was also a prominent theme in the text that accompanied Anna's quilting, of special note being the hunger winter[17] of 1944:

> It was very cold with a big shortage of food in the big cities. We received coupons for sugar, butter, and many other household items, and we had to queue for hours to get it from the stores. The food kitchens supplied soups. We called it *Goudvissensoep* [goldfish soup], as it was more or less water with a few slivers of carrots in it. My mother made pancakes from the pulp of sugarbeet and tulip bulbs.

Figure 24. Forest where V2s were launched (Anna)

Many elderly people and children died of malnutrition. A quarter of a million Dutch perished in the war, including 18,000 from starvation in the western provinces during the hunger winter of 1944–1945. At the height of this famine hundreds of thousands of Dutch became severely malnourished (Schulz).[18] These circumstances left deep scars of grief (Figures 25 and 26).

Johanna Binkhorst recalls,

> The winter of 1944–45 was the hardest, and was an exceptionally cold winter. There was not much fuel to warm our houses, gas was only available for a few short hours a day, and the food situation became critical. We all tried very hard to help one another and to make the best of a very difficult situation. We had to make do with surrogate tea and coffee, and clothes would be recycled, with a continuous effort to make new out of old.

Food drops are a common theme for the women from both the Netherland East Indies and the Netherlands. Three of the women prepared a quilting piece with the food-drop theme (Figures 27–29).

Figure 25. Searching for food amidst the bombing (Gerada)

Wilhelmina remembered the food drops in the Netherlands East Indies of peaches, chocolates, and corned beef. In the Netherlands, Gerada recalls food parcels with bread and chocolate being dropped by Allied planes: 'Oh, the joy of it. We thought the bread was cake and the chocolate was heavenly'.

A poignant memory for Anna Dijkman-Tetteroo was waiting '[i]n the streets with Dutch flags to welcome the drop of food. I'll never forget that I received a small loaf of Swedish white bread and was allowed to eat the whole loaf at once'. Frances recalled the 'thrill of being able to go for a walk in her street and the food parcels that discharged chocolates, sweets, powdered milk, and the butter and jam on real bread that followed'.

However, such thrills were short-lived as the Indonesian Revolution for Independence followed immediately after the war. The women were evacuated to the Netherlands as soon as the revolution began, so their quilting pieces reflect their wartime experiences, not the events related to the Indonesian Revolution for Independence 1945–1949.

War and post-war trauma had wreaked havoc with the lives of most young Dutch, whether they originated in the Netherlands or the Netherlands East Indies, hence their search for another homeland.

Figure 26. Queuing for food (Johanna)

Australia Quilt: Memories of Arrival

The women's reflections on immigration to Australia were the subject of the second quilt they tackled, but for the sake of historical chronology, I consider this work last. Three of the women's quilting pieces portray carefree Australian beach holidays (Figures 30 and 31).

Figure 27. Food drop—Netherlands (Anna)

Odyssey Quilt Two (Figures 30–32) sees the quilters engage with Australia's flora and fauna. This emphasizes the process discussed earlier about bonding with the new environment. Figure 32, by Frances Larder, depicts life in the bush, grass and eucalyptus trees and much loved Australian birds: parrots and a Kookaburra (Figures 33 and 34).

Around 5 per cent (500,000) of the Netherlands population left the country during the twenty-five years after the war to settle in nations around the world, including the United States, Australia, Canada, South Africa, Brazil, and Argentina (Peters 'Dutch Migration').

Some of the push-and-pull features that drove her family and also the wider community to move are discernible in the following extract from Johanna Binkhorst's artist's statement:

> After liberation from the Nazi Occupation, the rebuilding of our country began. However, extreme housing shortages and continuing rations made life difficult. We wished for a better life and in 1950 were tempted by advertisements for people to come to sunny Australia. With the promise of a fresh start, we packed up that year to begin a new life there with our daughter.

Figure 28. Food drop—Netherlands East Indies (Frances)

Background History

For Netherlands East Indies Dutch emigrants, repatriation to the Netherlands, where most had never been before, had been very disappointing. Most had come straight from Japanese internment camps and the life-threatening Indonesian Revolution for Independence that followed, only to find themselves made to feel unwelcome by the Netherlands Dutch. This is identified as the main push factor in their subsequent move to Australia.

Intending emigrants were enticed to Australia at information evenings and via posters, billboards, and fliers that depicted an attractive lifestyle of booming industry, boundless opportunity, full employment, good working conditions, and a chance to own your own vehicle and your own home filled with white goods. This level of material wealth was unheard of in the post-war Netherlands, Europe, or Britain. Moreover, it could be achieved with 'passage assistance' to which both receiving and relinquishing governments contributed (Peters *Milk and Honey*) (Figures 35 and 36).

Vera Rado's journey was unusual at the time:

> I flew to Sydney in December 1950 as a self-funded immigrant, then found a job and settled into the Australian way of life without any difficulties. I actually found it easier

Figure 29. Food drop—Netherlands (Johanna)

to assimilate into the Australian community than I did into the Dutch when I went to school there [in the Netherlands] for three years. In Australia, I worked as an office stenographer and secretary for almost thirty years. I married a Dutchman […] in 1953 and moved to the suburb of Smithfield where I still live today. I divorced him in 1979 and enrolled into Macquarie University to fulfil my ambition to study philosophy. I graduated with a BA in 1989.

Vera's visual diary portrays not only Australia's physical environment but also aspects of its history. An aircraft in flight signifies her entry into Australia, which is symbolised by a map of Australia showing that she settled for Sydney. Her drawing of a tree with very strong roots burrowing into the soil is the central image suggesting that Vera equates successful migration with putting down roots in her new homeland. The leaves of the tree are adorned with stamps and the caption 'assimilation'.

Her use of postage stamps to relate early history is clever; each stamp explicitly chosen because it visually narrates an important theme in Australia's past. The Tall Ships, Europeans coming ashore and taking the land for Britain, pastoral leases, the economy predicated as it was on the sheep's back, the exploration of Antarctica, ANZAC (Australian and New Zealand Army Corps of the First World War) and mateship, and the

Figure 30. Beach holidays (Francis Widitz)

Second World War are all there. Interestingly, Vera gave stories of Aboriginal culture the most space in her visual diary, implying their high level of importance for Australia's historical past. She also placed the caption 'naturalisation' under the postage stamp 'history' to confirm that, having absorbed Australia's history, she was now ready to become an Australian citizen. In addition, she has placed the caption 'integration' over the roots of the tree, possibly suggesting a deeper attachment than 'assimilation', given that 'one cannot become another'. For the Dutch, the term *aanpassen* (to accommodate, adapt) implies a more realistic resettlement strategy (Peters 'Just a Piece of Paper'; 'Expectations versus Reality').

Vera placed around the base of the tree visual images of personal achievements that she believes Australian society had facilitated, signifying that these had helped her put down roots in Australia. The indicators include photos and captions of her parents' home at Pott's Point Sydney, her rented home in Paddington, her first 'wholly owned' home in Smithfield, and her car (a Volkswagen sedan). They also include the proudly displayed Bachelor of Arts document that she received at Macquarie University in 1989.

Vera's quilting piece gives primacy to Australia's flora and fauna. However, it also notates the opportunities Australia afforded her family, as promised: a home of their own with car and furnishings, a job, and the opportunity to study. Vera compares Australia's climate with the cold and snow of Europe, and these she portrays as in a cloud in

Figure 31. Beach holidays (Gerada)

the sky—like a dream—no longer relevant to the life she is leading in Australia. Vera's quilting piece also illustrates some of the important issues her visual diary raises about being Australian, including the cultural meaning the land holds for Aboriginal people. Vera sums up her feelings as she looks back:

> My past experiences—many of them traumatic—have not prevented me from enjoying my retirement years. I am contented, having learnt not to take anything for granted and to be grateful for the 'pluses' in life.

Anna Dijkman-Tetteroo's visual diary incorporates a quotation from John Steinbeck—'Time is indeed a sacred gift and every day is a little life'—to make a point similar to Vera Rado's about the gift of life. Perhaps the violence of the war years had this impact.

In terms of the possibility of bonding with the new environment, it is worth noting that at least 50 per cent of Anna's visual diary, like Vera's, is centred on Australian Aboriginal culture, Australia's flora and fauna, and her feelings about these. Another third is given to visually illustrating the impact of European invasion and colonisation on Indigenous culture. In the remainder she has two personal photos on display: one of herself and

Figure 32. Life in the bush (Frances)

Figure 33. 'A new future'—Dutch emigration propaganda, 1950s

Figure 34. Australian immigration propaganda, 1950s[19]

her husband in front of their home and the other of their two dogs, representing (re)settlement and bonding in her Australian world (Figure 37).

Anna describes her impressions of Australia (Figure 38):

> I arrived in Australia, a new land, a different land, a terrifying land, in so many ways. I did not know what I was in for when I came across hundreds of sheep on the highway. But I came to love this land of extremes in due time, with its swathes of parched land that stretches across two-thirds of the country. I was mesmerized with the colours of the arid outback and the coastal regions in the north containing rain forests with many different kinds of tropical trees and flowers, which were different from the tulips and daffodils in Holland. It is difficult to describe the beauty of the Kimberley with its rugged ranges, spectacular gorges, waterfalls, wetlands, abundant wildlife, and above all the unexpected and unusual-shaped Boab trees, which look like they have their roots upside down.

Johanna Binkhorst's quilting piece has a theme central to post-war migration—building your own home while living in makeshift accommodations. An unexpected irony on arrival for many newcomers was Australia's critical building and building material shortage, which meant housing was as hard to come by in Australia as it had been in the Netherlands. Therefore, as soon as it was feasible, the majority of newcomers went in pursuit of accommodation so that they could leave the crowded migrant reception centres. However, finding a place to live proved to be the single greatest obstacle that 1950s migrants encountered. It was estimated that between

Figure 35. Arrival—visual diary (Vera)

250,000 and 300,000 new homes were needed Australia-wide. The situation was a consequence of the decline in building during the Depression and the Second World War and massive building material and skilled labour shortages (Creek). Migrants leaving migration camps moved into all sorts of makeshift accommodations: train and tram carriages, tents, back verandas, and even car crates or a room built from the crates in which proper kit homes had arrived that wealthier migrants or the state had imported.

Vicky van der Ley recalls the impact on her family of the shortage of accommodation in Australia:

> Three months [after arrival] … we moved to a block of land we had bought in Hay Street, Ryde. It took eight years to build our house, during which time we lived in a caravan. These were long and difficult years, and I had to do the cooking in the backyard over a gas primus. Our wonderful friends and neighbours were always very helpful and made this time much more enjoyable. That was fifty years ago, and in that time I have revisited both Holland and Indonesia—but Australia will always be my home.

In conclusion, in producing the Odyssey Quilts out of their combined knowledge and experience, the women were able to produce quilting pieces that communicated memories

Figure 36. Arrival quilting piece (Vera)

Figure 37. Visual diary—Impressions of Australia (Anna)

Figure 38. Building your own home (Johanna)

and emotions with which others of their cohort, and beyond, could identify. Their quilting pieces relating to childhood and also those depicting early impressions of Australia support the philosophies of Bender, Norberg-Schulz, and de Certeau about the role that bodily experiences play in creating strong bonds in relation to place. They also reveal the capacity that humans have to develop new bonds in places other than their country of origin.

The remarkable achievement of the Odyssey Quilts, when they are viewed in conjunction with the autobiographical stories the creators of the quilts have told in other ways, is that they capture artistically the experiences of a period in Dutch history that has also become an important part of the history of immigration and multiculturalism in Australia over the second half of the twentieth century in the aftermath of the Second World War.

Notes

[1] See https://collection.maas.museum/object/522933.
[2] Currently, over 335,000 Australians claim Dutch origins, the majority being descendants of the great post-war emigration wave—1949 to 1970—when some 170,000 Dutch migrated to Australia. Around 10,000 of these immigrants were originally residents of the Netherlands East Indies. See http://www.abs.gov.au/ausstats/abs@.nsf/Lookup/2071.0main+features902012-2013.
[3] The women held their first exhibition at the Casula Power-House Museum in Sydney in 2005. Since that time, the quilts have also been exhibited in Liverpool, New South Wales and at the Coffs Harbour Art Gallery and the Fremantle Arts Centre in Western Australia.
[4] Ivonne Chapman was also one of the quilters. Her main involvement in the Odyssey Quilt Project included providing workshops and ongoing guidance for the artists involved and providing Frances with assistance throughout the project.
[5] The women's oral history quotations in this chapter are derived from their artist biographies.
[6] The quilters' first names are generally used here in acknowledgment of the close bonds that developed between them and because it feels appropriate for the personal nature of the memories they share with each other and with the viewers of their work.
[7]

> Wayang, also spelled Wajang (Javanese: "shadow"), classical Javanese puppet drama that uses the shadows thrown by puppets manipulated by rods against a translucent screen lit from behind. Developed before the 10th century, the form had origins in the thalubomalata, the leather puppets of southern India. The art of shadow puppetry probably spread to Java with the spread of Hinduism.
>
> (*Britannica*)

[8] See http://www.timetravelturtle.com/2012/10/borobudur-temple-largest-buddhist-indonesia/.
[9] In the period 8 March 1942 to 15 August 1945, it is estimated that the Japanese interned as prisoners of war (POW) approximately 42,000 members of the Koninklijk Nederlandsch Indisch Leger and 22,000 Australian male military personnel. These Australians and Dutch were inevitably thrown together, particularly as working parties on the infamous Burma-Thailand Sumatra Railway. Some were also transported from POW camps as forced labour to Japanese timber, engineering, mining, construction, and many other projects around the Asia-Pacific region, and to Japan, where they had to work under deplorable and life-threatening conditions. By the end of the war, of these groups, some 8,000 Dutch and just over 8,000 Australian prisoners had died of ill treatment, starvation, and diseases such as yellow fever, malaria, and

cholera. Around 20 to 27 per cent of those in these camps died as a consequence of forced labour, malnutrition, and disease, including dysentery and typhoid. Vickers claims that

> over 100,000 ... civilians (including some Chinese) were put into detention camps on Java, while a further 80,000 military from Dutch, British, Australian and US Allied forces ended up in prisoner-of-war camps. The death rates in those camps ranged from 13 to 30 per cent.
> (Adrian Vickers, *A History of Modern Indonesia* [Cambridge UK, Cambridge University Press, 2005, 87])

[10] Johanna Binkhorst unpublished bio—Odyssey Quilts Project.

[11] https://www.documentatiegroep40-45.nl/dwangarbeid_oud/indexenglish.htm. In Europe, as the war progressed and Germany's fortunes began to deteriorate rapidly, more and more young Dutch were picked up and sent to work for the Nazi war machine. In due course, complete birth cohorts were bound to work. Ultimately the number rounded up to work as forced labour for the Nazi war machine would total over 475,000. More than 30,000 of them would perish through hunger, sickness, maltreatment, and acts of war.

[12] See https://www.verzetsmuseum.org/museum/nl/museum?gclid=EAIaIQobChMItLCzpd-32gIVqpztCh3eyAIWEAAYASAAEgJQe_D_BwE.

[13] See http://www.resilience.org/stories/2005-07-04/jatropha-oil-promising-clean-alternative-energy.

[14] On this quilting piece, the sun symbolises Japan's dominance, and the explosions symbolise the Japanese invasion.

[15] Adek camp at Sluisweg, now Jl. Tambak in Central Jakarta. See more at Khouw.

[16] Captain Sonei was sentenced to death by the Dutch War Crimes Court for his ill treatment and cruelty of Dutch and Australian internees and POWs in the Netherlands East Indies (The Argus [Melbourne], 7 September 1946, 4); Borch, F.L., Military Trials of War Criminals in the Netherlands East Indies 1946–1949, Oxford Univerity Press, 2027.

[17] http://www.environmentandsociety.org/tools/keywords/dutch-hunger-winter-1944-45.

[18] In the winter and spring of 1944, after a railway strike, the German occupation limited rations such that people, including pregnant women, in the western region of the Netherlands, including Amsterdam, received as little as 400 to 800 calories daily (https://www.verzetsmuseum.org/museum/en/tweede-wereldoorlog/kingdomofthenetherlands/thenetherlands/thenetherlands,june_1944_-_may_1945/the_hunger_winter). The amount of food available on ration dropped steadily. More than 20,000 people died of starvation. The transport of coal from the already liberated south also ceased. Gas and electricity were shut off. People chopped down trees and dismantled empty houses to get fuel.

[19] Australia Invites You, Courtesy NAA, C3939/1 N1957/75106 Pt. 2.

References

Bender, B. 'Introduction'. *Contested Landscapes—Movement, Exile, and Place*. Ed. B. Bender and M. Winer. London: Bloomsbury, 2001. 1–20. Print.

Booth, W. James. *Communities of Memory: On Witness, Identity, and Justice*. Ithaca, NY: Cornell UP, 2006. Print.

Britannica.com. Wanang. 2014. Web. 16 November 2015.

The Concealed History of the Netherlands East Indies, 1942–1949. Dir. Betty Naaijkens-Retel. 2014. Film.
Creek, M. 'You Know You've Got a Roof over Your Head: The War Service Homes Scheme'. *On the Homefront*. Ed. Jenny Gregory. Perth: UWA Publishing, 1996. 251–56. Print.
Das Bundesarchiv. 'Voluntary Forced Labourers? Expanding West'. 2010. Web. 16 November 2015.
De Bary, William T. *Sources of East Asian Tradition: The Modern Period*. New York: Columbia UP, 2008. Print.
De Certeau, M. *The Practice of Everyday Life*. Translated by Steven Rendell. Berkeley: U of California P, 1984. Print.
Gupta, A., and J. Ferguson. 'Beyond "Culture": Space, Identity, and the Politics of Difference'. *Cultural Anthropology* 7.1 (1992): 6–23. Print.
Helmrich, Hetty Naaijkens-Retel. *Buitenkampers* [documentary film]. 2013. https://www.vpro.nl/cinema/films/film_10677377_buitenkampers_.html.
Hofstede, E. W. *Thwarted Exodus*. Den Haag: Staatsuitgeverij, 1964. Print.
Khouw, Ida Indawati. 'Jakarta a "City of Hell" during Occupation'. *Jakarta Post*. 1 September 2001. Web. 16 November 2016.
Krancher, Jan. 'The Camps'. *The Indo Project*. 27 October 2011. Web. 16 November 2016.
Leach, N. 'Belonging: Towards a Theory of Identification with Space'. *Habitus: A Sense of Place*. Ed. J. Hillier and E. Rooksby. Aldershot: Ashgate, 2002. 281–95. Print.
Lynch, K. *The Image of the City*. Cambridge, MA: Technology Press and Harvard UP, 1960. Print.
The Macquarie Dictionary. North Ryde, NSW Macquarie Library, 1997. Print.
Norberg-Schulz, C. *Genius Loci: Towards a Phenomenology of Architecture*. New York: Rizzoli, 1979. Print.
Oosterman, Gordon, et al. *To Find a Better Life: Aspects of Dutch Immigration to Canada and the United States, 1920–1970*. Grand Rapids, MI: National Union of Christian Schools, 1975. Print.
Peters, N. 'The Dutch Migration to Australia: Sixty Years On'. It's Time to Burn the Wooden Shoes. Ed. M. Schrovner and M. van Faassen. *Tijdschrift voor Sociale en Economische Geschiedenis* 7.2 (2010). Print.
———. 'Expectations versus Reality: Postwar Dutch Migration to Australia in 400 Years of Dutch Connections with Australia'. *National Maritime Museum Conference Proceedings*. Ed. L. Shaw. Sydney: Australian National Maritime Museum, 2006. Print.
———. 'Going Dutch: 400 Years of Netherlanders in Australia'. *A Changing People: Diverse Contributions to the State of Western Australia*. Ed. R. Wilding and F. Tilbury. Perth, WA: Department of the Premier and Cabinet, Office of Multicultural Interests, 2004. n.p. Print.
———. 'Just a Piece of Paper: Dutch Women in Western Australia'. *Studies in Western Australian History* 21 (August 2000): 53–74. Print.
———. *Milk and Honey but No Gold*. Perth, WA: UWA P, 2001. Print.
———. 'No Place Like "Home": Experiences of the Netherlands East Indies as Real, Virtual, and Politically Contested Reality'. *Associated European Migration Institutions (AEMI) Journal* 12 (2014): 72–81. Print.
———. *From Tyranny to Freedom: Dutch Children from the Netherlands East Indies to Fairbridge Farm School, 1945–1946*. Perth, WA: Curtin U, Black Swan Press, 2008. Print.
Rado, Vera. *In Japanese Captivity: The Story of a Teenager in Wartime Java*. Sydney: Self-published, 2006.
———. 'Life Story.' 2008. Unpublished manuscript.
Schulz, Laura C. 'The Dutch Hunger Winter and the Development Origins of Health and Disease'. *Proceedings of the National Academy of Sciences* 107.39 (2010): 16757–58. Print.
Swarthmore.edu. Global Nonviolent Action Database. 'Dutch Citizens Resist Nazi Occupation, 1940–1945'. 2011. Web. 16 November 2015.
van Dulin, J., W. J. Krijsveld, H. G. Legemaate, H. A. M. Liesker, and G. Weijers. *Geillustreerde Atlas van de Japanese Kampen in Nederlands Indië*. Asia Minor: Ziedrikzee, 2002. Print.
Willemsen, Wim. 'Breaking Down the White Wall: The Dutch from Indonesia'. *The Dutch Down Under, 1606–2006*. Ed. Nonja Peters. Perth: UWA P, 2006. 132–49. Print.

Is Autobiographical Writing a Historical Document?: The Impact of Self-Censorship on Life Narratives

Magda Stroińska and Vikki Cecchetto

Self-censorship is the act of censoring one's own written or spoken words, usually out of fear of punishment or loss of face but sometimes also out of respect for the feelings of others. Self-censorship is usually related to public discourse—whether in literature, the visual arts or in the media—but it may infiltrate private discourse if the speaker has internalised the rules of what can be said and what should remain unsaid. In this chapter, we analyse the linguistic mechanisms of self-censorship in the context of the autobiographical writings of Andrzej Czcibor-Piotrowski (1931–2014). The self-censorship can be traced to the political system in works published prior to 1989 but a different justification is required for the novels published after 1989 when there was more freedom of expression in Poland. We believe that the explanation can be traced to the trauma experienced in childhood and the inability to disclose true memories, even some 70 years later. This finding challenges somewhat the concept of the documentary value of life writings from periods of historical terror.

To survive, you must tell stories. (Umberto Eco, The Island of the Day Before, [207])

If something is to stay in memory, it must be burned in: only that which never ceases to hurt stays in memory. (Friedrich Nietzsche, On the Genealogy of Morals II, S 3, [497])

Autobiographical accounts relating eyewitness testimonies of events, both written and oral, have become the most prolific genre of the twentieth and twenty-first centuries (Gilmore 128; see also Conway; E. Wiesel qtd. in Felman and Laub 5; Jensen and Jordan; Ahmed and Stacey, among others). The history of our age has been marked by an overwhelming number of violent conflicts, man-made tragedies, and natural disasters. Witnessed through the media by millions worldwide, the testimonies of those who experienced this trauma first-hand are eagerly sought by all, but for very different reasons: for the survivors, it is to tell the story in order to start their healing process; for the 'onlookers', it is to try to understand the reasons for what has happened.

The readers of such accounts tend to assume that, because they are autobiographical narratives, they should be treated as factual. As Leona Toker asserts,

> Narratology is helpless in determining the factographic status of a narrative, since all the features of a non-fictional autobiographical narration can be faked 'in cold blood': it is only sources external to the text that can corroborate its factographic status. (*Varlam*, section 1, par. 5)

This study argues that autobiographical accounts, especially involving some form of trauma as retold by authors themselves, cannot be assumed to be documentary in nature: they may be true or partially true, but they are not necessarily so (cf. Eakin; Toker *Varlam*; Toker 'Toward a Poetics'; Portelli, *The Battle*; Portelli *The Order*; McAdams). The accounts, as published, may be the result of a combination of state censorship in those countries where the government controls the version of history that is shared within and outside the nation (the external need), and self-censorship of the authors themselves who want to exert control over the retelling of their life story to an audience (the internal need).

How do censorship and self-censorship by the author distort the retelling of traumatic memories in an autobiographical document? Based on our previous research on trauma narratives, censorship, and propaganda, we analyse the linguistic manifestations of self-censorship in the context of autobiographical writings by Andrzej Czcibor-Piotrowski, Polish poet, translator, and writer, and two accounts of his childhood spent in deportation camps in Russia during World War II. In the novel published prior to 1989, *Prośba o Annę* (*A Plea for Anna*, 1962), his self-censorship can be traced to the political system, but a different explanation is required for the novels published afterwards. We demonstrate that the interpretation for self-censorship in his later novel, *Rzeczy nienasycone* (*Insatiable Things*, 1999) can be found in the trauma experienced in childhood and the inability to disclose his true memories, even some seventy years later: a finding that may challenge the concept of the documentary value of life writings.

Life Writing, Autobiographical Narratives or Egodocuments, and 'Truth'

The Greek etymology of the word 'autobiography' points to αὐτός (self) and βιογραφία, consisting of two parts: βίος (life) and γραφία (writing). In his *On Autobiography,* Philippe Lejeune has defined autobiography as 'the retrospective narrative in prose that someone makes of his own existence when he puts the principal accent upon his life, especially upon the story of his own personality' (4). Some earlier forms of autobiographical writings, usually associated with powerful or celebrated men (with only some examples of women's autobiography), include *apologia* or an attempt at 'self-justification', oration, confession, *historia*, *memorias* or memoirs, *vitae*, diaries, and journals. All of these genres are categories of testimonials—that is, first-person life narratives that can be used as tools for the construction of identity, what Jacques Presser has designated as 'egodocuments' (qtd. in Dekker 7).

Since the beginning of the twentieth century, autobiographical writing or life narratives in the Western tradition have been taken up by the ordinary person. One might ask, with Kim Lane Scheppele, 'Why is there such a rush to storytelling? Why has narrative become such an important format everywhere?' (2073). It seems that in our cultural accounts of our lives, especially vis-à-vis individual and collective traumatic events, an autobiographical narrative is considered an important and creditable eyewitness account.

The autobiography genre creates certain assumptions about the 'content' or information supplied by the writer. For example,

> *intentionality* signals the belief that the author is behind the text, controlling its meaning; the author becomes the guarantor of the *intentional* meaning or truth of the text, and reading a text therefore leads back to the author as origin. (Anderson 2)

It is natural for the reader to take a narrative assumed to be autobiographical to also be documentary—a truthful representation of the events described. And yet, as Roland Barthes put it, every author can say, 'It is my political right to be a subject which I must protect' (15). Paul Eakin indicates 'the misrepresentation of autobiographical and historical truth' as the first of the three primary 'transgressions' for which autobiographical writers have often been criticised (113–4). Eakin goes on to say, 'Definitions of autobiography as a literary genre inevitably feature truth-telling as a criterion' (115). In recalling the memory of a traumatic event by people who lived it, the author, the protagonist, and the subject are one, and what 'they' remember as the 'facts' might be different from the 'official' history, or from what really happened (since these two may not be the same either).

What Mechanisms of Self-Censorship Can Be Found in Autobiographical Writing?

Autobiographical writing or oral history interviews are subject to both external censorship and internal or self-censorship. They are also subject to the narrator's imperfect and evolving memory and shifting perspective. As Alessandro Portelli suggests, 'Oral history is a work of relationships; in the first place, a relationship between the past and the present, an effort to establish through memory and narrative what the past means to the present' ('What Makes' 21; see also Dekker 7–8; Anderson 17; van der Kolk and Fisler; McAdams). As such, an autobiography may be a moving *document humaine*, an egodocument, but cannot always be considered a faithful record of the events described. According to Portelli, life narratives are 'concerned with the interplay between what we can assume to have been fact and what happens in the realm of memory, including imagined events and false memories' ('Dialogical Relationship' 1; cf. Portelli, *The Order*, for a description of how 'fossilized lies' can become established in life writings/autobiographical narratives). The result is that what seemed an eyewitness account of historical facts retold by a survivor may have originated as an amalgam that incorporates second-hand accounts, 'cultural' mythology related to the event/story and 'true' historical elements.

In order to complete a person's life story, traumatic experiences need to be incorporated into a survivor's life narrative, but they are often too painful or shameful to be retold, relived, and worked through without some form of professional help and counselling. In many cases, though, the writer/interviewee may consider the medium at his or her disposal as 'professional help' to confront and deal with the still unresolved pain, guilt, and shame of events from decades earlier. This point is worth emphasising since there seems to be a marked trend especially in today's media reports to construct news and 'history' based on 'eyewitness' accounts of personal, political, social, or even natural disaster traumas. It is not yet clear if the trend to 'eyewitness testimony' reports comes as a result of the media's exploitation (obviously for the sensationalism or realism factor, but also in order to appear to the public at large as a more 'truthful' report of the event) of the willingness of people now to recount their traumatic experiences on camera, or whether this trend is in response to a search for the 'truth' behind an event. Nevertheless, it would be prudent to question the validity of such 'eyewitness' accounts because the 'truth' of the facts recounted by the witness may not always be certain.

The desire to externalise an 'eyewitness' account of a traumatic event may be constrained, though, by societal rules. Any society, whether ancient or modern, has both formalised as well as 'unwritten' customs, taboos, or laws by which speech, dress, religious observance, and sexual expression are regulated. Usually, a person or group is appointed or takes on the responsibility of overseeing the application of, and the adherence to, both the codified and the unwritten laws of behaviour by all members of the society, although there may be different 'interpretations' of the rules for certain sectors, for example, the group in power. In the service of the protection of the three pillars—family, church, and state—censorship is combined with political power to become a form of symbolic violence over individuals or groups, which results in the limitation on, or the impediment of, the free communication of thoughts, ideas, and behaviours. As soon as a person externalises one's thoughts and ideas in written, oral, or visual form, that person immediately become visible to those in power and can become accountable to these more or less formalised rules. The result may be that these thoughts and behaviour can be targeted for 'modification' (i.e., 'punishment') by the authorities. For this reason a resistance develops in many individuals to disclosure, and there is a conscious effort to keep 'sensitive' things hidden in their works (Splichal).

Some individuals may choose to self-censor out of fear of being punished by external forces if they say or do something that goes against what is deemed the public good. Others may choose to self-censor because they consider some topics off-limits, as being either damaging to their own identity, offensive or damaging to others, or simply inappropriate for the time, place, or audience. In autobiographical writing where traumatic situations are recounted, an author may choose to silence or modify in some way the facts associated with a memory that may be inconvenient, painful, shameful, or dangerous. In some cases, the speaker or writer may create a diversionary problem to alert the hearer or reader to a traumatic event or experience but in such a way that the real trauma is never touched upon or disclosed. We claim that this may have been the case in the writings of Andrzej Czcibor-Piotrowski.

What remains unsaid or is consciously or voluntarily silenced by the author are the inconvenient events or people who should not be mentioned in public speech. For example in Czcibor-Piotrowski's first novel published in Communist Poland, the Soviet invasion of Poland in 1939 and the deportation of Poles to Central Russia are never mentioned. This version of the novel is conditioned by the self-censorship of certain facts or memories resulting from external pressures—for example, the political censorship existing at the time, which the writer had internalised. But self-censorship of another kind is also operant in his novel. Traumatic events that survivors do not want to talk about or acknowledge, sometimes even to themselves, may be edited out of their eyewitness accounts. This action represents a form of denial, common to those who suffer or have suffered from posttraumatic stress disorder (PTSD). If they decide to disclose information relating to the personal trauma, the information may be disguised by changing the facts. Another phenomenon related to PTSD is dissociation, the use of magic and derealisation. The lack of chronological plot development in the 1962 version of Czcibor-Piotrowski's novel reflecting the flashback nature of traumatic memories, and the use of magic and dreamlike scenes—forms of derealisation and depersonification—are characteristic of a survivor dealing with traumatic experiences (APA).

Traumatic memories recalled in autobiographical narratives provide more than just purported documentary knowledge of historical events. Often they offer an alternative insight into the author's inner world and are therefore significant for understanding the traumatic experience and its aftermath, as well as helping us to understand the role of memory and its lapses. They undeniably play a role in coming to terms with—or denying and repressing—the past (Morris-King). In addition, they are not open to criticism by others regarding the experience as it is recalled and relived by the survivor in the present (see LaCapra 86–113; *Trauma*).

Our Case Study: Two Autobiographical Novels by Czcibor-Piotrowski

Historical Facts about the Time Frame of Czcibor-Piotrowski's Novels

Czcibor-Piotrowski was born in 1931 in Lwów (then part of Poland), now Lviv, Ukraine. Together with his parents and older brother, he moved to Warsaw in 1933, but the family continued to spend most of its summers visiting the grandparents in Lwów or the vicinity or, from 1939, at the little estate purchased by Andrzej's mother, Dr. Wanda Zienkiewicz-Piotrowska, a gifted ophthalmologist. In August 1939, with the war approaching, Andrzej's father, Eugeniusz, a medical doctor, was called to active duty in the Polish Army and returned to Warsaw. The rest of the family was still on holiday near Lwów when, on 1 September, the Germans began the invasion of western Poland. At that moment, Lwów seemed a safer place than Warsaw, which was the target of German airstrikes. However, on 17 September, the Soviet Army invaded Poland from the east and occupied eastern Poland, along the lines defined in a secret treaty, popularly known as the Ribbentrop-Molotov pact, signed in the summer of 1939 by the foreign ministers of Germany and the Soviet Union. Polish families living in Lwów were subsequently considered enemies of the Soviet Union. Andrzej's father, serving in the Polish Army at the eastern front, was taken prisoner by the Soviet troops and only escaped death with the help of a fellow doctor working for the Russians. Having fled the POW camp, he found his way to Lwów to see his family and then went back to Warsaw where he obtained travel documents for his wife and both sons so that they could legally—and safely—get back to German-occupied Warsaw.

While it is difficult to know whether the actions of the author's mother were really due to her father's death and the illness of her mother, or because of an alleged romantic involvement (all occurring in Lwów at the time), Andrzej's mother took the boys to the border but, once there, tore up the documents, turned around, and returned to Lwów. Soon after, in 1940, most Poles still in Lwów were deported to Central Russia. The family was taken to a small village, Panino, where the mother worked in a cantina for lumberjacks until the acknowledgement of her medical qualifications resulted in her being moved to a makeshift hospital to work. There she contracted typhoid fever and died in July 1941. The older son Renek, who had developed polio, had been sent to a larger town for treatment before his mother's death, thanks to her medical connections. Therefore, at her death, nine-year-old Andrzej was left alone and was able to survive only thanks to the help of certain Polish and Russian families in the village.

On 22 July 1941 Germany invaded the Soviet Union, thus nullifying the Ribbentrop-Molotov treaty. This suddenly changed the political situation of Polish prisoners of war and deported civilians: their Polish citizenship was restored, and the surviving civilians were allowed (by the Russians) to join the newly formed Polish armed forces and leave the Soviet Union for Persia. Andrzej was one of the children who were part of the civilian contingent following the Polish Military in the West (loyal to the Polish government-in-exile) which was travelling from Russia through Persia, arriving finally in Scotland to join other Polish military groups there. His brother Renek, who had meanwhile been adopted by a Russian family after the death of his mother and his release from hospital, returned to Warsaw in 1945 at war's end to join his father. Andrzej, located by the Red Cross in Scotland, returned to Poland in 1947. Both boys wanted to study medicine to follow in their parents' footsteps. However, because professional clans were discouraged under the Communist rule, Renek became an engineer and Andrzej studied languages, becoming a translator of Czech, Slovak, Russian, and English literature and a published poet.

How Does Czcibor-Piotrowski's Trauma Affect His Autobiographical Novels?

All these experiences undeniably had a traumatising effect on the childhood and adolescence of Czcibor-Piotrowski, who, because of his mother's death and the forced absence of his father, was left without a nurturing parent with whom to talk and with whom to work through any psychological consequences of his traumatic experiences. As Levine and Kline state,

> [C]hildhood traumatic symptoms can continue to show up in the months, years, and sometimes decades after the incident. The sooner the child is given first aid by a familiar adult or treated by a professional when indicated, the less likely it is that secondary symptoms will develop. Another certainty is that lasting symptoms tend to be pervasive. (69)

There are two versions of Czcibor-Piotrowski's retelling of his traumatic experiences during the Second World War. In 1962, he published a short novel, *Prośba o Annę* [*A Plea for Anna*], subtitled *Opowieść w szesnastu snach* [*A Tale in Sixteen Dreams*], recounting his experiences in Russia. This second title could already be seen as an initial form of self-censorship in order to mislead the Communist regime censors, as if to highlight that this is the recounting of something that happened in his mind only and not necessarily what he really experienced. Because of the censorship imposed during the Communist regime, the book never specifies what had actually happened to him and to his family. Then, in 1999, he published a much longer work titled *Rzeczy nienasycone* [*Insatiable Things*], the first volume of a trilogy dealing with the full period of his wartime experiences, including his deportation, his adventures in travelling from Russia to Persia and finally his experiences after reaching Scotland. Both novels analysed here deal with Czcibor-Piotrowski's childhood memories from the period of forced deportation from Lwów to Central Russia, but the differences between them are considerable.

The first novel, *Prośba o Annę*, the 1962 version of the events, has third-person narration and a protagonist who has different names at different stages of life. All the names, though, seem to refer to the same person, being various nicknames for Wiktor ('Victor')—a name with possible significance. The story has a nonlinear plot development, with events that happened several years apart mixed together through the process of association. Neither the physical location for most of the events nor the time frame is ever indicated. The only place name mentioned is the idealised place of Wiktor's happy childhood—Czciborów, a name based on Piotrowski's middle name, Czcibor, which he later added officially to his surname.

The reasons for the complex temporal structure and the changing identity and name of the protagonist may be twofold. In the first place, the book describes the painful memories of a child separated from his family, the death of a beloved mother, the suffering and loneliness and the recurring memories of traumatic events—all within the historical context of the deportation of Polish civilians by Soviet authorities. But according to the political propaganda existing in Poland in the early 1960s, the Soviet invasion and occupation of eastern Poland never happened, and neither did the deportations. Suggesting otherwise would have represented career suicide for any historian or writer. The internalised rules of political censorship of the time translated into the self-censorship of the author, making him disguise the factual elements of the memoir and forcing him to obscure references to places and events, not only in the novels referring to this period, but also in his poetry. For example, in the poem titled 'Matka' [Mother] (Świadectwa 17–18), written in 1971 on what would have been the thirtieth anniversary of his mother's death, we encounter obvious elements of self-censorship:

[1] The author alludes to the death of his mother, to the fact that she does not have a marked grave, and that her sons, as the only remaining relatives, could never mourn her where she died and is buried.
[2] Where she died is not specified or named, except to say that her ashes are 'Underneath rye / Underneath forest / There in the north'.
[3] The year of her death is alluded to by reference to himself: 'it was my tenth year'. Since the poem was published in 1971 and he was born in 1931, at the time of publication it would have been the thirtieth anniversary of her death. The reference to dying in 1941, in the north, would be enough information for Polish readers to infer that she had been one of the deportees to Siberia.

More than 30 years after the publication of the first novel, Czcibor-Piotrowski felt the need to revisit his childhood memories, resulting in the publication of the 1999 novel *Rzeczy nienasycone* [*Insatiable Things*]. He explains it as follows:

> After many years, I was overtaken by the temptation to write prose and, from the stories 'We' and 'Notes from Memory' published in a London newsletter, this book started to slowly weave itself into being. I swam against the current of merciless time into the barely remembered past. And then, all of a sudden, all that had happened almost 60 years ago returned. I was finding myself and those close to me again, Poles and others. I was recalling the awe and wonder, fears and hopes of childhood. I made this journey on a flying carpet and on a dragon: I experienced a marvellous feeling of vertigo. (*Insatiable Things* back cover notes; trans. by Magda Stroińska)

This second novel of his childhood memories differs from the 1962 version in many ways. It has first-person narration, making it clear that the narrator is the author (same name and enough autobiographical details to suggest the link). The story has a linear development, starting in August 1939 during the summer holiday at Rudniki, the mother's little estate near Lwów. The author/narrator is presented as a 'stable' actor, but with unstable gender identity: a little boy whose mother had wanted him to be a girl and whom she sometimes dressed as one until he started school. Places and dates are clearly identified: the reader knows that the book is telling the story of the deportation of Poles by the Soviet authorities in

1940. The collapse of Communist rule in Poland in 1989 made it possible for Czcibor-Piotrowski to situate his story in its proper historical context.

The self-censorship in Czcibor-Piotrowski's writing can certainly be traced to the political system prior to 1989. The 1999 novel, however, may not be telling the full story either, and the self-censorship in it requires a totally different interpretation. We believe that the explanation is in the trauma experienced in childhood and the inability to disclose true memories, even after 70 years, which may challenge somewhat the concept of the documentary value of life writings from periods of historical terror. What were the reasons for the boy's trauma, beyond the obvious effects of deportation and the loss of a parent? Czcibor-Piotrowski always refers to his mother as 'my beautiful and wise mother' or 'my beautiful and good mother', thus stressing his very strong emotional attachment to her. Similarly, he also talks about his mostly absent father as 'my handsome, wise, and brave father'. However, did these wonderful parents really do all that they could have done to protect their children from harm?

Since Andrzej had an older brother Renek, his mother had likely wished the second child to be a girl. When Andrzej was born, her disappointment may have been somewhat alleviated by the fact that he looked a lot like her: curly hair, delicate facial features, and a girl-like figure. She let him grow his hair long, put ribbons in it, made him wear dresses, and called him 'Uta'—an abbreviation from 'Laluta', a little doll. He may have interpreted this reaction to his being a boy as a form of maternal rejection, all of which must have been confusing for the child who was not yet sure about his gender identity and sexuality. To show the narrator's shifting feelings about his gendered self, Czcibor-Piotrowski alternates between feminine and masculine gender in the verbs, thus giving the reader an insight into his unstable personality. The following fragment illustrates this technique:

> ale pewnego dnia—a właśnie zaczynały się wakacje—urwało się to nagle i wcale nie wiem na pewno, czy <u>czułam się</u> szczęśliwa, kiedy zaprowadzono mnie do fryzjera, który długo nie chciał uwierzyć, że jestem chłopcem, ale w końcu uległ namowom, obciął moje śliczne długie loki [...], a gdy <u>spojrzałem</u> w lustro, już nie w sukience, ale w koszuli z wykładanym a la Słowacki kołnierzem i w krótkich spodenkach<u>, rozpłakałem się,</u> bo <u>wydałem się</u> sobie—<u>przywrócony</u> chłopięctwu. (*Rzeczy nienasycone* 16)
>
> *and then one day—the summer holidays were just about to begin—all of this came to an abrupt end and I do not know for sure whether I felt$_{FEM}$ happy$_{FEM}$ when they took me to a barber, who, for a long while, could not believe that I was a boy but finally was persuaded to cut off my beautiful curls [...], and when I looked$_{MASC}$ into the mirror, no longer in a dress but rather a shirt with a wide collar à la Słowacki and shorts, I started$_{MASC}$ to cry because it appeared$_{MASC}$ to me that I had been brought$_{MASC}$ back to boyhood.* [trans. by Magda Stroińska]

The 'beautiful and wise mother' was also the person who endangered her children's lives because of a romantic involvement. She decided to stay in the Soviet-occupied territory even though her husband had provided her and the two children with the necessary documents that would have allowed them to return to

Warsaw. While the sons may have been unaware of the larger picture and their mother's decisions at the time, when writing his books Czcibor-Piotrowski did know the full story. Thus, referring to his mother as 'beautiful and wise' may be more the way he saw her as a child and how he remembered her rather than what he really thought of her in later life.

As a writer Andrzej Czcibor-Piotrowski preferred to be viewed as an aging man obsessed by sexual issues to the point of pornography, rather than a child victim of sexual assaults and traumas. As an adult he tries to reclaim control over his life story—control that was taken away from him when he was a child and when he was taken advantage of by those who had more power. A young and sensitive boy, orphaned in the middle of a vast enemy country during the war, he was the helpless object of personal and historical traumas and most likely a vulnerable victim of abuse and assault, both in Russia and later when he travelled with the army, stayed in Persia, and went to school in Scotland. Looking like a girl, with his unstable sexuality and his insatiable need for love, he must have been easy prey for both peers and adults. In *Prośba o Annę* [*A Plea for Anna*], the opening scene of Book One (*Prośba o Annę* is divided into four 'books', each with internal chapters) includes an accusation by an unspecified court against the protagonist: 'You are charged with playing horses' (11). In the scene that follows, a child tells his mother that he was playing horses and the mother dismisses it, saying, 'Yes, that's the game with reins and a whip. Now, go away. Daddy and I are busy' (11). Only later, in a scene where the children are watching mating horses, is the reader made to understand that 'horseplay' is the child's word for sex. Thus, sex is the underlying motive of the book, and along with a child's innocent discovery of sexuality likely came something more sinister: children's exploitation by adults. It is also possible that little Andrzej had tried to tell his parents, but they had dismissed it or did not understand the significance of what the child was saying. Thus, again, they failed to protect him from harm, either by not understanding what he was trying to say or maybe by simply not being there for him (the father was away and the mother died and left him alone).

Normally, when recalling past experiences, childhood memories are filtered through the eyes of the adult who has emerged from them. This does not seem to happen in the case of Czcibor-Piotrowski's writings. In his chosen way of presenting his childhood experiences, the author has finally assumed his own agency and volition. A close reading of both of Czcibor-Piotrowski's novels is consistent, we believe, with our hypothesis that he was more traumatised than he was ready to admit, which led to their evident self-censorship.

Conclusion

What, then, is the impact of trauma on autobiographical narrative? First, when the life writing involves the narrative of a trauma survivor, the factual portrayal of the person's life events may be constrained by memory lapses or identity-protecting criteria: the memory of the traumatic event(s) may not reflect the historical facts or the truth about the events known by those not directly involved. Autobiography therefore should not be treated solely as a historical document. Trauma causes various degrees of self-censorship and leads the author to change or omit certain facts in an attempt to justify their actions or inactions at the time of the traumatic event, an attempt to gain agency. Such an approach is also an attempt to portray events and actions in such a way as to make them integrateable into the author's own life narrative, while allowing them to continue to be acceptable to the perception others have of them. Czcibor-Piotrowski was one of the last of the WWII deportees who wrote about the trauma he experienced during that period without explicitly acknowledging his trauma. In this way he embodied the concept of the 'wounded storyteller' (Frank). An autobiographical narrative, then, is often a face-saving act, as the reported memories mask those that may be too painful to be revealed. We readers should take with a grain of salt the seemingly factual nature of autobiographical writing, especially by trauma survivors.

Postscript

Sadly, Andrzej Czcibor-Piotrowski passed away on 16 May 2014 after a long illness, before we could share the final version of our chapter with him. We had been fortunate in receiving very encouraging feedback from him on an earlier version. After reading it, he suggested to Magda Stroińska that he would welcome her writing his biography. We took this as a positive endorsement of our analysis.

References

Ahmed, Sara, and Jackie Stacey. 'Testimonial Cultures: An Introduction.' *Cultural Values* 5.1 (2001): 1–6.
Anderson, Linda. *Autobiography*. New York: Routledge, 2001.
APA. *Diagnostic and Statistical Manual of Mental Disorders DSM-IV-TR (Text Revision)*. Washington, DC: American Psychiatric Association, 2000.
Barthes, Roland. *Camera Lucida: Reflections on Photography*. Trans. Richard Howard. New York: Hill, 1982.
Conway, Jill Ker. *When Memory Speaks: Reflections on Autobiography*. New York: Vintage Books, 1999.
Czcibor-Piotrowski, Andrzej. *Prośba o Annę [A Plea for Anna]*. Warsaw: PAX, 1962.
Czcibor-Piotrowski, Andrzej. *Rzeczy nienasycone [Insatiable Things]*. Warsaw: W.A.B., 1998/99.
Dekker, Rudolf, ed. *Egodocuments and History: Autobiographical Writing in Its Social Context since the Middle Ages*. Hilversum: Verloren, 2002.
Eakin, Paul John. 'Breaking Rules: The Consequences of Self-Narration'. *Biography* 24.1 (Winter 2001): 113–127.
Eco, Umberto. *The Island of the Day Before*. Trans. William Weaver. San Diego: Harcourt, Brace & Company, 1995.
Felman, Shoshona, and Dori Laub. *Testimony: Crisis of Witnessing in Literature, Psychoanalysis, and History*. New York: Routledge, 1992.

Frank, Arthur W. *The Wounded Storyteller: Body, Illness, and Ethics*. Chicago: U of Chicago P, 1995.

Gilmore, Leigh. 'Limit-Cases: Trauma, Self-Representation, and the Jurisdictions of Identity'. *Biography* 24.1 (Winter 2001): 128–39.

Jensen, Meg, and Jane Jordan, eds. *Life Writing: The Spirit of the Age and the State of the Art*. Newcastle upon Tyne, UK: Cambridge Scholars, 2009.

LaCapra, Dominik. *Writing History, Writing Trauma*. Baltimore, MD: Johns Hopkins UP, 2001.

Lejeune, Philippe. *On Autobiography*. Ed. Paul John Eakin. Trans. Katherine Leary. Minneapolis: U of Minnesota P, 1989.

Levine, Peter A., and Maggie Kline. *Trauma Through a Child's Eyes*. Berkeley, CA: North Atlantic, 2007.

McAdams, Dan. *The Stories We Live By: Personal Myths and the Making of the Self*. New York: Guilford, 1993.

Morris-King, Shelley. 'Eliciting First-Person Accounts of Childhood Wartime Experiences and Perceived Impact on Psychological Well-Being'. *Journal of Aggression, Conflict, and Peace Research* 1.3 (2009): 48–57.

Nietzsche, Friedrich. "On the Genealogy of Morals." Trans. Walter Kaufmann. *The Basic Writings of Nietzsche*. New York, Toronto: Random House Inc., 2000.

Piotrowski, Andrzej. *Świadectwa*. Warsaw: PAX, 1971.

Portelli, Alessandro. *The Battle of Valle Giulia: Oral History and the Art of Dialogue*. Madison: U of Wisconsin P, 1997.

Portelli, Alessandro. *The Order Has Been Carried Out: History, Memory, and Meaning of a Nazi Massacre in Rome*. New York: Palgrave Macmillan, 2003.

Portelli, Alessandro. 'A Dialogical Relationship: An Approach to Oral History'. *Expressions Annual* (2005). 7 Dec. 2014. <http://swaraj.org/shikshantar/expressions_portelli.pdf>.

Portelli, Alessandro. 'What Makes Oral History Different'? *Oral History, Oral Culture, and Italian Americans*. Ed. Luisa Del Giudice. New York: Palgrave Macmillan, 2009. 21–30.

Scheppele, Kim Lane. 'Telling Stories'. *Michigan Law Review* 87 (1989): 2073–98.

Splichal, Slavko. 'Manufacturing the (In)visible: Power to Communicate, Power to Silence'. *Communication and Critical/Cultural Studies* 3.2 (2006): 95–115.

Toker, Leona. *Varlam Shalamov*. 'Testimony and Doubt: Shalamov's "How It Began" and "Handwriting"'. 2007. 15 August 2012. <http://shalamov.ru/en/research/121/>.

Toker, Leona. 'Toward a Poetics of Documentary Prose—from the Perspective of Gulag Testimonies'. *Poetics Today* 18 (1997): 187–22.

Trauma in the Twenty-First Century. Spec. issue of *Life Writing* 5.1–2 (2008).

van der Kolk, Bessel, and Rita Fisler. 'Dissociation and the Fragmentary Nature of Traumatic Memories: Overview and Exploratory Study'. *Journal of Traumatic Stress* 8:4 (1995): 505–25.

Material Memory and the Digital

Paul Longley Arthur

Over the past two decades, memory, understood as both the act of remembering and a means of storing memories, has been relocating itself. In its daily usage it has been moving from the mind to the computer—from neurological systems to digital technologies—as people increasingly outsource memory to digital devices. In this essay I focus on the changing nature of remembering—and forgetting—in the digital era. With an emphasis on personal stories I ask: How is intergenerational memory transfer changing as a result of digital media technologies? Specifically, what are the implications of the shift to digital storage and communication processes for the way we retain, pass on, or receive private and intimate material? How has this changed the way we see ourselves and view our lives, and allow others to see ourselves and our lives?

In the second decade of the twenty-first century, we rely more than ever on computer memory to enhance our human capabilities to remember and recall information, both institutional and personal. This digital extension of memory is enabled by the capture and preservation of vast amounts of data that form exponentially expanding archives. Whereas in the past it was common practice for individuals to carefully keep in boxes or drawers documents, photographs, and mementos—to be perused from time to time or worked through later by family members—personal information is now being stored on multiple computer hard drives, portable media, and cloud storage systems. In a remarkably short period of time, the term 'memory' itself has changed. Over the past two decades, memory, understood as both the act of remembering and a means of storing memories, has been relocating itself. In its daily usage it has been moving from the mind to the computer—from neurological systems to digital technologies—as people increasingly 'outsource' memory to digital devices.[1]

In this essay I focus on the changing nature of remembering—and forgetting—in the digital era. With an emphasis on personal stories I ask: How is intergenerational memory transfer changing as a result of digital media technologies? Specifically, what are the implications of the shift to digital storage and communication processes for the way we retain, pass on, or receive private and intimate material? How has this changed the way we see ourselves and view our lives, and allow others to see ourselves and our lives? I explore these questions in the context of a great loss for which digital technologies have been responsible, at the same time that they have delivered marvellous benefits. The benefits have been so vast and so dazzling that this loss appears to have been accepted with little comment and, it sometimes appears, barely noticed. However, in this transitional time, as the last pre-digital generation ages, the impacts on history, biography, and life writing are coming sharply into view. The loss that concerns me here is the rapid disappearance of the material objects that were at the heart of memory transmission. In private and intimate human stories, such objects often carried a depth of significance, emotional power, and authority that the digital world cannot match. In the wider context of the vulnerability of all digital data—through erasure, failure of archiving systems, or inaccessibility as a result of technological obsolescence—this may seem like a minor matter. On the other hand, it may represent one of the biggest shifts in recorded history in the way memories are transmitted within families and communities.[2]

Google vice president Vinton Cerf, known as one of the fathers of today's Internet, has

> warned that we risk becoming a "forgotten generation" or even that this will be "a forgotten century" because as computer hardware and software become more quickly obsolete, we are losing information. Data are disappearing into a digital black hole, despite the fact that we supposedly have more access to books, letters, and photographs than ever before. (Cooper, 'Thanks')

This warning points to a massive and very recent change in the way we select the things we want to remember—or forget—as well as the way we store and share memories.

Using digital cameras as the example, Anthony Funnell makes a similar point: 'Photographs have a way of transporting us back to the time they were taken and dredging up hidden memories. However, new research shows that in our rush to digitally record every moment we're actually remembering less and less.' Funnell draws attention to 'people who are capturing media images that they never, ever view'. He thinks of these as 'orphaned memories'. He argues that 'we have in some senses offset our memory into these devices'.[3] As a result there is a global proliferation of graveyards of abandoned images that are never visited and therefore no longer used for their traditional purpose of releasing and sharing memories.

In 1936, when Walter Benjamin was foreshadowing the consequences of the shift to 'mechanical' reproduction of art, he was presenting ideas that were applicable not only to works of art but also to other kinds of objects that would traditionally have had special significance (219–53). The techniques of reproduction, he believed, would detach 'the reproduced object from the domain of tradition' (223). They would also disconnect it from 'the history to which it was subject throughout the time of its existence', including 'the changes it may have suffered in physical condition over the years as well as the various changes in its ownership' (222). Thus, in Benjamin's words, to attempt reproduction though these technical means would be 'to pry an object from its shell, to destroy its "aura"' (225).[4] He regarded early photographic portraits, however, as a special case:

> It is no accident that the portrait was the focal point of early photography. The cult of remembrance of loved ones, absent or dead, offers a last refuge for the cult value of the picture. For the last time, the aura emanates from the early photographs in the fleeting expression of a human face. (228)

Following the trajectory of Benjamin's predictions, photographs in the digital world have lost their aura. And in our increasingly online world, other memory objects that have traditionally connected us to the 'absent or dead' are disappearing. Most strikingly, letters and diaries of the traditional kind are almost a thing of the past. Personal stories and memories are now generated differently and housed elsewhere. After many centuries of progressively greater exteriorisation of memory, digital technologies have suddenly taken the process to new extremes and into a different arena. Most notably, memory is moving away from material objects into undifferentiated space.[5]

In the digital era we no longer write onto inscribable surfaces. Today's tablets are not made of stone, and the words we write on them disappear into a void, a cloud, that is everywhere but nowhere. The effect is that the externalisation that began in writing, and was enabled by print (Le Goff vi–vii), has come back on itself. Words are no longer tethered to material carriers. Physical writing surfaces are being digitised so as to provide a preserved ethereal surrogate for the real and tangible. All this means that memory, individual and collective, is no longer encased in the way it has for so long been; memory is being set loose from its conventional containers and vessels. We send our most intimate thoughts and writings out to online services, and they can be difficult to re-find or bring back in.

For centuries, letters—private, handwritten, and with no copies—have played a central role in the sharing of life experiences and in the keeping and transmission of memory between generations. This applies to the messages they carry and also to the materials on which they are written. Other kinds of material objects have also served as guardians of memory, as 'containers and vessels', literally or metaphorically.

For more than 20 years, my Ukrainian grandfather, Petro Olijnyk, wrote letters from Adelaide, Australia—where he had lived since his arrival as a Displaced Person in 1949—to his brothers Pavel and Gregori, who lived in two different regions of Ukraine. Kept in a box in my mother's attic are an assortment of smaller containers with more than 100 letters from Pavel, and a similar number from Gregori and other members of the family. Pavel died in the late 1980s. Gregori continued to write, but with declining health and failing memory, he passed on the task to his son Dmitri, who faithfully continued the correspondence right up to the time of Petro's death in 2005. Separated during World War II, the brothers were unable to make contact with Petro until the early 1970s. Because he never saw them again, Petro depended entirely on the exchange of letters to share details of his life with his brothers and to learn of all that had happened to them since the traumatic time of separation. Through them he received news of his mother, whom he had long assumed to have died at the start of the war. She was still living in his childhood village and was able to send messages via her sons. For my grandfather, all this gave the letters an emotional power far beyond the information they contained. The letters were eagerly awaited, and opening them was a ceremonial occasion followed by readings at the kitchen table—with translation, explanation, and historical elaboration—for anyone who was willing to listen.

Letters from the younger brother, Pavel, were stored in an old cylindrical tin that once would have contained canned food of some kind, probably Polish pickled cucumbers. Piles of other letters are squashed down neatly into a variety of boxes: a 'Monople Magnums' plywood cigar box dated May 1956; a pre-decimal 'Cadbury's Roses' tin that once contained one pound of chocolates, its lid decorated with a picture of a spaniel; half a dozen small cardboard boxes that each contained 100 sheets of 'Ilford' five-by-seven-inch photographic paper—or loosely bundled by long-since-withered elastic bands. Each box triggers family memories: my grandmother collecting empty cigar boxes from the job she had in the cigarette store at the Adelaide Railway Station soon after arrival as a refugee, her older son's passion for photography, my grandfather's resistance to throwing anything away. The boxes are all clearly labelled in my grandfather's writing with a black marker.[6] Here and there, the '*par avion*' envelopes—with their blue and red borders, USSR or Russian stamps, and foreign handwriting, show signs of silverfish activity as lacework along their edges. When my grandfather died a decade ago I took a photograph of his tin of Pavel's letters, feeling that it represented a vital link with his family history and also with mine (Figure 1).

As recently as when this correspondence with family in Ukraine began, it would have been an expectation that the letters would be handed down, translated, and kept as an archive, not only of a family, but of an era, providing insights into the nature of international communication between people who had been displaced and intimate details of daily life in Soviet Ukraine. Untranslated and enclosed in their envelopes, the letters are nevertheless overflowing with memories—of my grandfather reading them aloud, translating as best he could and explaining historical contexts and cultural differences. Once, when Pavel's pig produced a litter, he was proud to report that the family had named one of the piglets Kateryna after my mother. While there was much amusement at this dubious cross-cultural honour at the Australian end, there was also a warm sense of connection through a family story that will continue to be passed down. Lying dormant in their containers, these letters represent a treasure chest of stories, but they are also material objects that can be touched; in handling them, one literally feels a connection with another historical world. That they were handled in that other world gives them a presence that connects lives across

Figure 1. Pavel's letters stored in a tin (photograph by the author)

generations and continents. In them are many private moments between brothers, to do with their mother, family, tragedy, the brother who was imprisoned (although it is not clear why), the daughter on my grandmother's side who died. There is something highly evocative about their materiality: the rusty tin bursting with letters, the stained boxes, and my grandfather's handwritten labels. The letters themselves carry the tangible traces of the people who wrote them—in the paper, the ink, the smudges, the script, the stamps, the notes in margins, and the folds. For my grandfather, as he read and reread and replied to them, the letters would have generated what Virginia Woolf once called 'the drag of the face on the other side of the page' (102) that letter writers feel. This element is rapidly disappearing as we entrust our communication and our memories to digital repositories.

Even though the Ukrainian script is indecipherable to me, the tin contains the key to memories that can be accessed through human rather than technological means. This is not to say that scanning and storing them digitally is not necessary or helpful, but rather that the physical presence of the tin and its contents underlines the general point made by naturalist and photographer Paul Evans with reference to the digital world, that 'there's a huge gulf [...] widening between the image and the thing itself' (qtd. in Funnell). The digital surrogate cannot replicate the effects of the memory object's location in time and space—in Benjamin's words, 'its unique existence at the place where it happens to be' (222).

Knowledge has been 'handed down' or 'passed on' in many ways through history, and 'memory is the raw material of history'.[7] Technologies at any time allow representation and transfer to take particular forms. In a less globalised world, transfer was more culturally specific than it is now. There were (and are) primarily oral traditions, where knowledge has been communicated in song or dance or through objects, and the idea of cultural transmission is closely bound up with notions of ritual and its power.[8] Ritual and tradition created stability and enduring 'truths' that, in the case of Indigenous or First Nations peoples, provided a link across sometimes tens of thousands of years of history. In the expression 'handed down', there is the suggestion of something manual—of an object moving from the hands of one human being to another. But the stories and knowledge that flow from powerful and evocative objects do not stay static; they change and gain new life as they are engaged with and passed on.[9]

In what is sometimes now called the 'analogue' era, there were limits on what you could keep. You could not send 100 letters in a day. There were natural physical limits on what one person could produce and what could be collected and stored. Today, because online storage spaces are practically infinite, at least to a typical user, filling them up is almost impossible. The tendency is to either throw material away or save it until later—to sort at another time—but there is no incentive to sort. There is no urgency to identify the most significant items. There is no need to find a tin or a box to keep things in. The items do not have to be contained. And so stuff accumulates, or depending upon one's nature, it might be discarded. Memory is not deposited into a physical place in the way that it used to be and cannot be retrieved in the same way. Because memories no longer come packaged in discrete parcels—as letters, photo albums, or postage parcels, for example—they are not entered as defined spaces to be explored but rather brushed against in chance encounters within systems of open-ended connectivity. This is not to say that containment constricts the act of remembering; it is likely that the reverse is true. Memory can be unleashed all the more powerfully by the very fact that its artefacts exist as a limited package that has survived the passing of time—has been 'spared'.[10] Helene Cixous's extended meditation on memory and forgetting, *Double Oblivion of the Ourang-Outang*, dramatises this recognition by both initiating and deferring a painful emotional engagement with a forgotten box, 'full of *old stuff*', stored at the back of a wardrobe, that is referred to as 'a collection of things happened fifty years ago' (58):

> The Box is full of time. What is in the Box: a cube of time, carved up into the matter-history of my life, and structured like corrugated cardboard: two sheets of paper, with a third sheet—also corrugated—in between. There is no more

wonderfully durable material than paper time. [...] I am fascinated by the textual properties of life on paper, by the fact that memory has found a way to protect itself and to protect the youth of its characters in the folds of a fabric! (71)

In the text the box takes on a kind of life and becomes a central 'character' that has 'cunningly' entered the narrator's life: 'I dream [...] about the Box character. [...] I say: "the box, he" ... rather than "she" or "it"' (14). With its 'profusion of pasts' waiting to be released, the box in Cixous's story is a powerful metaphor for material memory. However determinedly the writer tries to resist the lure of the box, 'It could not escape me. I could not escape it' (3). In his *Meditation on a Pair of Shoes*, Paul Claudel also points to the lifelike quality of memory-laden things:

> Ordinary objects [...] take on a sort of personality, their own face, I could almost say a soul, and the folklore of all nations is full of these beings more human than humans, because they owe their existence to people and, awakened by their contact, take on their own life and autonomous activities, a sort of latent and fantastic wilfulness. (1243)

Cixous's cardboard box is like the betel bags in Janet Hoskins's account of traditional Kodi storytelling in that it is an 'ordinary object' filled with stories.[11] The Kodi stories, however, were ready to be shared. Cixous's are so private that the writer is reluctant to reveal them even to herself:

> The Box still somewhat given over to Secrecy. I could have moved away. I still could. Throw this Box away without looking at it. Thrust the Subject down so deep into the cupboard's throat that not a word will ever be spoken about it again. There is still a little time. A little later I would start to forget it. Later still I will have forgotten the forgetting. All might yet be lost. (39)

Cixous's narrative dramatises the enormous power that a collection of objects from one's past can wield. Their digital surrogates would not have the same effect. A passionate collector, Benjamin celebrated 'the thing world'[12] represented by personal archives and all kinds of material objects—and recognised their value in the domain of memories.[13] In the past such objects had to be physically contained and bounded, and their location—at the back of a cupboard, in an attic, in a box or drawer for safekeeping—was an integral part of their material 'life'. Typically, such items were selected to form a miscellany that represented significant moments, relationships, or stages in one's life, or they were things that it seemed wrong to throw away, even if the reasons weren't always clear.

Now, because we have access to so much information that accumulates at such a fast rate, we no longer select and sort. In our throwaway consumer society, nothing has much value; even our personal data exceeds the amount we can reasonably address. By keeping everything, we ultimately value nothing. This condition is disorienting and dispossessing. We do not have self-possession in the sense of a set of possessions that define us. Nor can we rely on the familiar habits and rules of narrative to build a sense of identity. Our memory repertoire—its configuration, location, and mechanisms—are being affected by this shift. Anna Poletti and Julie Rak sum up the situation in this way:

> What we are, who we love, how we live and communicate with others, how we think of our life histories, even if we make our lives as stories at all, and even what being alive means: these are all states of being increasingly mediated through online digital environments. (20)

Digital information accrues around us whether we like it or not. Emails are never permanently deleted in the way one might have shredded or burnt a sensitive letter. They remain on servers. Incriminating Tweets, once released into the digital ether, cannot be reined back in even when the user deletes them. Information has a life of its own and continues to circulate ad infinitum. The recent ruling in the 'The Right to Be Forgotten' case allows, for the first time, a user to request that web pages be removed from search indexes so that it is harder to find them—but they are not destroyed ('EU Court').

More than ever, ongoing selectivity is crucial to building any collection. Libraries and archives have changed their policies to discard aggressively, because they have to. There is simply too much to keep. A national library's mandate to collect one of every published book seemed to be a manageable, bounded task when publication was a gatekeeper to quality, meaning that not everyone could be a published author. Now publication formats and opportunities have multiplied to the extent that anyone *can* be an author. Open-access principles mean that works are sometimes published at no cost to the user. The task of deciding what is worthy of keeping online is therefore challenging, especially for national institutions. The National Library of Australia's Pandora project is a web archive project that takes snapshots of web content over time, for selected sites. It resembles the Internet Archive's global Wayback Machine, which has the goal of recording a much wider array of web content, but the snapshots recorded are just that. They only go so many page layers deep, often do not preserve dynamic content or even images in certain formats, and take only a static 'print' view of pages. In other words, these archives are poor representations of the original, yet this process shows that even when websites are closed and their servers switched off, erasing their traces is impossible.

Personal archives present their own range of unique challenges for preservation. So much personal data is being produced, self-generated, and automated around us. Because we can keep everything, nothing seems to have much value. On my to-do list I have 'must sort photographs'—because I can never find what I care about. However, if one's personal information is not adequately backed up or synced with a cloud storage service, then it is very vulnerable. Once personal data enters the realm of the Internet, erasing or hiding it is almost impossible, but if one keeps it 'safe' on a private hard drive or device, then it faces other risks such as hardware failure, software obsolescence, accidental deletion, or loss through many other common hazards. These digital documents are even more vulnerable if they are scanned copies of originals that have since been discarded. Major archives have data preservation, format migration, and management plans in place, but for individuals the data can be lost in an instant. We speak about the digital 'dark ages' as being the 1990s, but much personal data faces exactly the same risks today. In some sense the risk is amplified because we rely so heavily on digital storage and there are so many born-digital documents. However diligent people may be, they can lose their data and their archives. What's left, then, for many of the current generation is a Facebook or other social media archive—in which people present themselves very differently from the way they would reveal themselves in private diaries or letters.[14] Social media interactions, whether the material presented is of a private nature or not, are public performances whose repercussions cannot be known or contained.[15] As Helen Lewis has said, in a recent article, in the 'public squares' of the Internet, 'no one knows who is really talking to whom, and—surprise!—a conversation between anything from two to 2,000 people can feel disorienting and cacophonous' (Lewis).

The question that arises is: Where and in what form we will find records of intimate transfer in the future? Could it be that we need to accept the demise of the significant material object and accept its replication via the digital? If so, it will be necessary to manage excess—shedding it, casting it off, locking it out. In other words, old patterns of prioritisation may need to be reactivated to build a culture of selection, via archiving tools designed for private material but with 'keys' available, like keys to chests that held secret documents in the past.

What we may finally need to give up and forget is the endless capacity to remember. Data wealth is now commonly experienced as data overload, data ambush, and communication chaos. We may also need to accept that while the computer has not yet provided a satisfactory alternative to individual handing down of memories, it has the capability to build a perpetual living archive that can be designed to serve the purpose of memory transfer more effectively than it has to date. But once again, the need for selection and containment looms large.

In a recent book titled *Delete: The Virtue of Forgetting in the Digital Age*, Viktor Mayer-Schönberger cites Stacy Snyder's work as a reminder of the importance of 'societal forgetting'. By 'erasing external memories', he writes, 'our society accepts that human beings evolve over time, that we have the capacity to learn from past experiences and adjust our behaviour'. In traditional societies, where mistakes are observed but not necessarily recorded, the limits of human memory ensure that people's sins are eventually forgotten. By contrast, Mayer-Schönberger notes, a society in which everything is recorded 'will forever tether us to all our past actions, making it impossible, in practice, to escape them'.[16]

However, at the individual and societal levels, there are things that need to be remembered and transmitted. In a medium known for its inherent instability and dynamic flux of formats that need to continue to adapt to meet software and hardware demands, imagining that digital technologies can provide a safe and secure mode of capturing and preserving threatened or delicate cultural memory between generations would be unrealistically optimistic. By contrast, a letter or written diary can sit literally for generations, without being tended to, and can still be found. The locked casket, used by Gaston Bachelard, Benjamin, and Cixous as a metaphor for the safekeeping of intimate secrets, remains a potent symbol of the necessity to forget, so that we can contain and bring to the present moment and retain for the future the things that can unlock what we want to remember:

> The casket contains the things that are *unforgettable*, unforgettable for us, but also unforgettable for those to whom we are going to give our treasures. Here that past, the present and a future are condensed. Thus the casket is memory of what is immemorial. (84)

Paradoxically in this digital age, which promises permanent memory and recollection and information access from any time and any place at the press of button or swipe of a screen, forgetting and erasure of personal digital effects and data may be more important than their collection in order to preserve a memory vault of private and intimate information that can be passed on. In Marc Augé's words, 'We must forget in order to remain present, forget in order not to die…' (89)

Notes

[1] 'That's Linda Henkel, a professor of psychology at Fairfield University in Connecticut in the United States. […] [H]er recent research suggests that the way we use digital cameras acts as a form of outsourcing of memory' (Funnell). All Funnell quotations in this essays are from this transcript.

[2] 'But a moment's pause makes it startlingly clear that the way we pass information on—and connect past and future generations—is changing radically' (Cooper 'Memories' 17).

[3] 'It's the notion that when you take pictures of things you are kind of counting on the camera to remember for you, and because of that you don't really engage in the additional processing that would help you consolidate that experience into a more detailed memory' (Henkel qtd. in Funnell).

[4] As Benjamin explains, aura is related to uniqueness, and 'what is really jeopardized when the historical testimony is affected is the authority of the object' (223).

[5] 'While purpose-designed for writing on, these too are also physical objects that are carriers of the memories they contain. The ability to write down, or on, material surfaces or objects, had an externalising function, and this was greatly amplified by the appearance of printing' (Le Goff vi–vii).

[6] Amongst the letters from Ukraine lie a sprinkling of other family letters, some from later generations, as well as birthday cards, an old bank book, a pension card, postcards—an assortment of documentary bits and pieces.

[7] 'Memory is the raw material of history. Whether mental, oral or written, it is the living source from which historians draw. Because its workings are usually unconscious, it is in reality more dangerously subject to manipulation by time and by societies given to reflection than the discipline of history itself. Moreover, the discipline of history nourishes memory in turn, and enters into the great dialectical process of memory and forgetting experience by individuals and societies. The

historian must be there to render an account of these memories and of what is forgotten, to transform them into something that can be conceived, to make them knowable. To privilege memory excessively is to sink into the unconquerable flow of time' (Le Goff vi–vii).

[8] See Hoskins 56–57.
[9] See Hoskins 25–27.
[10] 'At this point I notice the pencil's *endurance*. Its stroke has kept a freshness that no ink could have preserved. Perhaps it is this untouched youth that touches me at first. Between the pencil and the Box, a certain harmony—a magical way of dispensing with time. *Of remaining spared*' (Cixous 108, emphasis added).
[11] See Hoskins 25–58.
[12] See 'Physiognomy of the Thingworld' (Ursula et al. 73–75).
[13] Benjamin's attitude to his collection of postcards is an example: 'There are people who think they find the key to their destinies in heredity, others in horoscopes, others again in education. For my part, I believe that I should gain numerous insights into my later life from my collection of picture postcards, if I were to leaf through it again today. The main contributor to this collection was my maternal grandmother' (qtd. in Ursula 171).
[14] A recent survey suggests that more than 70% of Facebook users present untruthful representations of themselves.
[15] 'The constant directive to "share" personal information on social media sites such as Facebook or Linked-In is an example of a media affordance, which asks for users to create a specific type of identity, one that can be shared' (Poletti and Rak 4)
[16] See also 'The Web Means the End of Forgetting'.

References

Augé, Marc. *Oblivion*. Minneapolis: U of Minnesota P, 2004.
Bachelard, Gaston. *The Poetics of Space*. USA: Beacon, 1969.
Benjamin. Walter. 'The Work of Art in the Age of Mechanical Reproduction'. In *Illuminations*. Ed. Hannah Arendt. Trans. Harry Zohn. London: Collins/Fontana, 1973. 219–53.
Cixous, Helene. *Double Oblivion of the Ourang-Outang*. Trans. Suzanne Dow, with the collaboration of Lucy Garnier. Cambridge: Polity, 2013.
Claudel, Paul. *Meditation on a Pair of Shoes*. Prose Works. Bibliotheque de la Pleiade. 1965.
Cooper, Glenda. 'Memories Lost in Digital Age'. *West Australian* 17 Feb. 2015: 17.
Cooper, Glenda. 'Thanks to Changing Technology, the Family Album Is Fading Fast'. *Telegraph.com*. 14 Feb. 2015.
'EU Court Backs "Right to Be Forgotten" in Google Case'. *BBC.com*. 13 May 2014.
Funnell, Antony. 'Digital Photography: Message and Memory'. *Future Tense*. Australian Broadcasting Commission. 18 Jan. 2015. Radio.
Hoskins, Janet. *Biographical Objects: How Things Tell the Stories of People's Lives*. New York: Routledge, 1998.
Le Goff, Jacques. *History and Memory*. Trans. Steven Rendall and Elizabeth Claman. New York: Columbia UP, 1992.
Lewis, Helen. 'Context Collapse'. *TheGuardian.com*. 1 Jan. 2015.
Mayer-Schönberger. Viktor, *Delete: The Virtue of Forgetting in the Digital Age*. Princeton: Princeton UP, 2009.
Poletti, Anna, and Julia Rak, Eds. *Identity Technologies: Constructing the Self Online*. Madison, Wisconsin: U of Wisconsin P, 2014.
'The Web Means the End of Forgetting'. *Ubuntuforums.org*. 5 Mar. 2015.

Ursula Marx, Gudrun Schwarz, Michael Shwarz, and Erdmut Wizisla, Eds. 'Physiognomy of the Thingworld'. *Walter Benjamin's Archive: Images, Texts, Signs*. Trans. Esther Leslie. London: Verso, 2007. 73–77.

Woolf, Virginia. *Collected Essays, Vol.1*. Ed. Leonard Woolf. London: Chatto, 1966.

REFLECTIONS

Because it's Your Country: Death and its Meanings in West Arnhem Land

Martin Thomas

The morgue in Gunbalanya holds no more than half a dozen corpses—and as usual it was full. So when the Old Man died in the Wet Season of 2012, they had to fly him to Darwin, only to discover that the morgue there was already overcrowded. So they moved him again, this time to Katherine, where they put him on ice until the funeral. The hot climate notwithstanding, things can move at glacial speed in the Northern Territory where the wags tell you that NT stands for 'Not today, not tomorrow'. The big departure had stalked and yet eluded the Old Man in recent years. Now he would wait six months for his burial. Only then would he be properly 'finished up', as they say in Gunbalanya, a place rich in many things: poverty, and euphemisms for death, among them.

The Old Man's name is indelibly associated with Gunbalanya, a settlement on the western border of the great Aboriginal reserve of Arnhem Land. But it is a name that cannot be used. In northern and central Australia it is customary, as a sign of respect for the dead and those who loved them, to avoid uttering the name of the deceased. The taboo on naming is a survival from a wider process of purging that traditionally occurred at the time of death. In earlier times—and sometimes even today—the possessions of the deceased were quickly burned or in other ways disposed of; the initial rites of grieving, involving preliminary treatment of the body, were rapidly performed; the site of death was abandoned. The transition to a more sedentary lifestyle means that it is no longer so simple to shift camp, while the high mortality rate, compounded with the financial impost and logistical complexities of assembling the necessary mourners, often from a wide catchment, has resulted in an almost farcical backlog of funerals. All of this contributes to the interminable prolongation of 'sorry business'. Nobody likes it, but like plenty of things that nobody likes, no one knows how to change it.

When people die, their house and the other places they regularly frequented are treated with the pungent smoke of green ironwood leaves. The persons close to them, and those who handled them after death, are also smoked. Smouldering boughs are brushed against the torsos of these mourners, or alternatively they stand near a fire and fumes are wafted about them. The purpose of these rites is to decontaminate the living from their contact with the dead and to protect them against the spirit who, at this transitional moment, might be confused, angry, mischievous or worse. The great changes of the past century have done little to diminish the potency of the spiritual realm, from which a person emerges at the time of conception and to which they return when mortal life has ended.

The conventions require that I do not name directly the Old Man whose last great work is the subject here. Formally, he is being referred to by the 'death name' chosen by his relations: Na-godjok Nayinggul. But here I will refer to him by the more familiar 'Wamud'. This is 'Bininj way'—the way we would speak of him if we were in Gunbalanya now. *Bininj* is the word for 'man' in the Kunwinjku dialect of the local language, Bininj Kunwok. The Aboriginal people of the west Arnhem region use it as a generic descriptor for themselves. Wamud was the Old Man's 'skin' name and I addressed him as such even when he was alive. This is customary among Bininj who generally avoid using a person's name in their presence, preferring instead terms of address that affirm the kin relationships that bind everyone together. Wamud often called me 'Bulanj'. That is my skin name, given to me by another old man, Kodjok, when he made me his classificatory son. That was his skin name, and I must refer to him as such, for just a week before Wamud's long-awaited funeral, Kodjok himself 'finished up'.

The names Wamud, Kodjok and Bulanj are not personal to those who are addressed as such. Each name signifies membership of one of eight classificatory groups to which everyone in the community belongs. Skin names descend through the maternal line and are used from the time of birth. People are expected to marry outside their own group into one of two other prescribed groups. While anthropologists, when they first encountered these sorts of classification, tended to emphasise their rigidity, the kinship system is in reality highly adaptive, as is evident in the practice of giving skin names to outsiders who are thereby positioned in the local taxonomy and, it is hoped, integrated to some degree into the network of reciprocal obligations that are the core of Bininj sociality.

The taboo on naming extends to representations of the deceased person, including photographs, films and voice recordings. Contrary to popular belief, the taboo is temporary. As time passes and mourners become reconciled to their loss, the person's name and image are permitted to resurface. How long this takes varies considerably. The age and status of the deceased have great bearing on the matter. In Wamud's case it will be some years until the people of Gunbalanya speak of him again by his 'true' name. He was a pre-eminent figure and an acknowledged leader, not only among Bininj but among Balanda—as we, the white people, are known in Arnhem Land. When speaking (or writing) there is no restriction on telling stories about the man whom we call Wamud: it's the *naming* of him that's the problem. Even at his funeral, or at the Christian part of it (held in an overflowing Anglican Church in Gunbalanya in June 2012), extended eulogies were delivered and the memories encouraged to flow.

My contact with Wamud, which began in 2006, was all to do with memories and their preservation. Historic films and sound recordings, made in Gunbalanya in 1948, were the impetus for my visiting the area. Equipped with digital copies of these films and records, which were originally intended for anthropological study, I hoped to interpret them with the true experts: their traditional owners. Produced during a post-war extravaganza known as the American-Australian Scientific Expedition to Arnhem Land, the films and recordings had been archived in collections in Sydney and Canberra. While study of them marked the beginning of our acquaintance, Wamud and I became friends more recently when we became entangled in some filmmaking of our own. The footage that we filmed together I can revisit privately, but it cannot be made public until the mourning period has ended. Inevitably, the footage that we shot shapes my impressions of him, even as I try to separate the diminished although still marvellously lucid figure recorded by the camera from the charismatic and physically robust individual whose eyes sparkled with astonishment when, on that first meeting, he was asked if he would like to see some film of the initiation ceremony known as the Wubarr.

Back in 2006, as a new chum to northern Australia, I was fortunate in having as a guide and tutor the linguist Murray Garde, a fluent Kunwinjku speaker, who is intimately versed in cultural protocols. Ceremonial knowledge in Arnhem Land is deeply gendered and the Wubarr is exclusive to men. To escape the gaze of unauthorised eyes, we retreated to the edge of town and set up my laptop in the hut of another old man, a distinguished painter, who is now also finished up. He and Wamud were close to tears by the end of the film. Much of the emotional

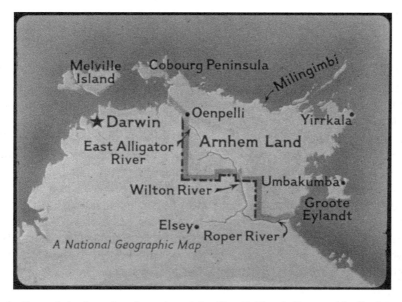

Figure 1. Map of Arnhem Land produced by the National Geographic Society for the American-Australian Scientific Expedition to Arnhem Land, 1948. Gunbalanya is identified by its mission name, Oenpelli. By permission of the National Library of Australia (NLA).

impact of the documentation is due to the breakdown of localised ceremonial traditions, of which the Wubarr was a key example. For reasons not easily explained, performances of Wubarr became ever more occasional from the 1950s. By the 70s it was defunct. This was not because male initiation had died out, as it has elsewhere in Australia. In west Arnhem Land it continues to this day. Instead, Wubarr was displaced by the pan-regional Kunapipi ceremony, a very different rite, dedicated to the Rainbow Serpent, which would emerge as the dominant mode of initiation. For Wamud, watching the old footage rekindled memories of the exquisite dance and music at the heart of the ceremony, and of his old teachers, finished up long ago, who had inducted him into the rite.

After Wamud had seen the footage he dictated a statement in Kinwinjku that Murray has since translated, asking that he be given access to this and kindred documentation:

> Today I have been shown the movie of the Wubarr ceremony taken a very long time ago, and so I wanted very much to see this. I was a young boy at the time and so I would like you very much to be able to make it [an edited copy] for me, so that in quiet times when I am not working, I'll go back to my camp and watch the images. It will make me reflect on all I know about that Wubarr ceremony and how at that time it was close to disappearing. Today that ceremony is gone, it lies buried forever. And so seeing that movie brought it back to mind again. Who can we find today who knows about that ceremony? A few of us know, but those old people are all gone now ... I would like to hold on to it. Myself and the other senior men here, we would share it together and gather together to watch it. We will watch to see how all those years ago, we were initiated by those old people into that Wubarr ceremony. If we don't see this film again, we won't be able to remember. Maybe all we would have is a name. The ... Wubarr ceremony has come alive again in those images they made. That is all.[1]

*

Four years passed, and my excavation of the archives generated by the 1948 expedition continued. A trip to Washington yielded a significant cache of films and photographs—of Wubarr and many other aspects of the Bininj world. In 2010, when Wamud expressed interest in making a film, it was clear to him, as it was to all with eyes in their head, that he had little hope of seeing it in its finished state. He had cancer, diabetes and more. As he admitted candidly, the beer, cigarettes and rough living had taken their inevitable toll.

We all spoke of him as the Old Man—this being respectful in a society where advancement in age confers prestige rather than irrelevance. But Wamud was hardly old by many people's standards. In fact I doubt he was 70. It is hard to be certain, for he was born in the bush where births went unrecorded. He was a baby when his family came in to Gunbalanya, then known to the wider world as Oenpelli Mission. Of course, by the standards of his own community Wamud *was* an old man. Indigenous life expectancy in the Northern Territory is even bleaker than it is in the rest of Australia. Territorian males who are Aboriginal can expect to live 61.5 years, compared to 75.7 for their non-Aboriginal contemporaries (Australian Bureau of Statistics).

With his raft of health problems, Wamud was heavily medicated and suffering at times excruciating pain. Cruel indignities had been inflicted on his once towering form, perennially crowned by a black cowboy hat. The hat remained while its wearer shrank. His attenuated physique, skeletal in its wheelchair, formed a surreal contrast to his face, lightly whiskered, devilishly handsome, and still exuding the playful quizzicality of earlier years. Infirmities notwithstanding, Wamud inhabited his body with the same easy confidence that he must have borne as a mature, initiated man in his physical prime. An operation had reduced his right foot to an ugly stump, but he made no attempt to hide it.

Old Wamud didn't wear a watch but he knew that he was racing time when he poured what he could of his precious energy into making the film. His motives for doing so were in a way obvious. He wanted to leave a record for future generations. I guess he knew that the footage would be suppressed during the mourning period and that eventually it would resurface. Characteristically, he was never a passive subject during our filming sessions, and I now realise that he was thinking very laterally about the type of record he was leaving of himself. This is apparent in the hours of interview that are presently embargoed, although it is more overt in other footage, filmed at his instruction, in which he doesn't appear.

On the first day of what would be our last shoot with him, in July 2011, Wamud asked to be taken through his ancestral country, west of Gunbalanya, to the East Alligator River. I was driving with him in the passenger seat. Adis Hondo, the cinema-photographer with whom I collaborate, and Suze Houghton, our sound recordist, sat in the back. At the causeway known as Cahills Crossing, Wamud

Figure 2. Cahills Crossing, East Alligator River. Digital photograph by Adis Hondo (taken at the request of Wamud).

asked Adis to get out and film the water. The river is the border between Kakadu National Park (which Wamud had helped establish) and the vast patchwork of Aboriginal homelands that is Arnhem Land. You can make what you like of Wamud's directive that we film the river. Paying homage to 'this beautiful country', as he often called it—picturing, singing and in any way honouring it—was always paramount. Clearly, filming the country was a way of marking his territory. But the more I look at the footage—the vegetation moving subtly with the breeze; the rush of water through the frame—it acquires a portent that I never perceived when I stood with Adis by the tripod, and not only because the river, almost *too* obviously, is a symbol of life.

Although located some 50 kilometres from the coast, the East Alligator is tidally affected at Cahills Crossing. With the assault of an incoming tide, the river promptly changes direction and within minutes the formerly innocuous causeway becomes impassable. A channel that flows in both directions, a saltwater phenomenon in freshwater country, it recalls the belief, common to various Aboriginal cultures of the northern coastline, that the recurrent transition between life and death, between mortal and spiritual existence, is manifest in the ebb and flow of the tides. It is not, I think, so very far-fetched to find a portrait as well as a landscape in that footage of the ever-moving river, a site of ebb and flow in which death as much as life is a creative force. As I look at that footage, which during Wamud's lifetime I would have been inclined to use for nothing more than the odd cutaway in our planned documentary, I find it suffused with his presence, although he sits outside the frame.

One year earlier he pointed out a shallow sandstone overhang where human bones lay piled in the dirt. He asked that we film them, and later spoke of them not as bones but as a 'man in his country'. This is relevant, for it was the restoration to their country of just such men and women that constituted Wamud's last great work.

*

Outwardly dishevelled and chaotic, the larger Aboriginal settlements like Gunbalanya are usually composed of a network of smaller communities, referred to as 'camps'. Kodjok lived in Argaluk Camp, so-called for its proximity to a boulder-strewn rise, a locale for ceremonies, called Argaluk. Wamud lived in Banyan Camp, named after a banyan tree of special significance. The designation 'camp' suggests that even now, more than a century after the pioneer buffalo shooter Paddy Cahill began farming and developing the site that would later fall into the hands of missionaries, the town has a transitory or provisional status, though for many occupants this is more a dream than a reality. Few Bininj own vehicles so it is difficult to travel to Darwin, to get to distant ceremonies, or to the small camps or clusters of huts that constitute the outstations (tiny settlements on often isolated homelands). Numbering 1171 inhabitants according to the 2011 census, Gunbalanya is a large settlement by Arnhem Land standards and it offers some employment and other attractions: two shops, a meatworks, an art centre, a clinic, a school and—controversially—a 'sports and social' club, licensed to sell alcohol. Wamud was a regular at this institution, and took pride in having played a role in its establishment.

Wamud's shared his house with his wife (who has passed away and cannot be named) and an indeterminate number of younger relatives. It was a metal construction, green in colour, no more or less dilapidated than those around it. The carcass of a four-wheel drive had made of the yard its final resting place. When his energy allowed it, I worked with him on the verandah during my visits. It was there that he and I watched a movie, copied from the National Geographic Society film archive, which shows the bones being stolen (Walker). That Aboriginal human remains were fetishised and collected in the name of science, and often taken to institutions in distant lands, is hardly breaking news. But the idea of bone taking being captured on a National Geographic film production, and in colour no less, shocks me even now—and I have watched it a hundred times. Like other inconvenient realities, ranging from land theft to massacre, there is a tendency to compartmentalise bone taking as a predilection of the nineteenth century when, as the film shows, it was nothing of the sort.

We first see the bone taker reaching into the hollow of a crevice; his rear end, protruding towards the camera, is stained with channels of sweat. Then, turning to face us, he unwraps a jawbone from a blackened shred of rag—evidence, one would think, that it is hardly of great antiquity. Bespectacled, and with lips pursed beneath a trim moustache, his officer's deportment is upset by a slash of blue headband that gives him a piratical craziness. He adds the jaw to a wooden crate already full of arm and leg bones, butted up against a skull.

Captured on celluloid, projected onto the surface that is the present, the theft is possessed of a luminous immortality. The bone taker's diary, which I have

Figure 3. The bone taker Frank Setzler holding a pistol and magpie goose. Photographer unknown. By permission of National Anthropological Archives, Smithsonian Institution (Photo Lot 36 Oenpelli_127).

read closely, makes it plain that the presence of the photographer was prearranged (Setzler, 28 October 1948). Built into the act of seizure was the attempt to transcend its fleetingness. Perhaps in an attempt to give it scientific import, the archaeologist and the cameraman made of the theft a pedagogical performance. We can see the archaeologist lifting and handling the skull he has taken, pointing out distinctive features, and then slotting the jaw in place and presenting to the camera its largely toothless grin. The final seconds of the sequence show a glimpse of scenery from the top of Injalak hill, which, for any local, is immediately identifiable as the site of the theft. There is a body of water, a yellowed strip of land, a few scant dwellings. We look, of course, at an earlier incarnation of Gunbalanya where, some 60 years later, Wamud and I sat on his verandah, grimly watching the spectacle on my laptop. The footage ends when the archaeologist and an assistant (a white man not seen until now), cross the frame and clamber downhill. Manhandling the now lidded crate, they disappear from the picture.

The diary records that the film was taken on 28 October 1948. The cameraman was *National Geographic Magazine* photographer Howell Walker and the archaeologist was Frank M. Setzler, Head Curator of Anthropology in the United States National Museum (now the National Museum of Natural History), a division of the Smithsonian Institution in Washington DC. They were visiting the region as members of the American-Australian Scientific Expedition to Arnhem Land. Just five days before Setzler took the bones, he and his fellow researchers had enjoyed the considerable honour of being admitted to the Wubarr ceremony, thereby producing the film and sound recordings that so enthralled Wamud and other old men.

The expedition was a large-scale research venture involving 15 men and two women that roved through Arnhem Land for much of 1948.[2] Somewhat propagandist in intent, its chief supporter within the Chifley government was the America-friendly Arthur Calwell, who, before he acquired the Immigration portfolio, served as Minister of Information. He was a great supporter of the Adelaide photographer-cum-ethnologist Charles P. Mountford, who led the expedition. Mountford had been lecturing in the United States in the last year of the war when he came to the attention of the National Geographic Society and the Smithsonian Institution, both flagship institutions. He persuaded them to sign up with the Commonwealth of Australia to become official partners in the expedition, a seven-month, anthropological, natural-history, photographic and filmmaking adventure, intended as an overt display of bilateral friendliness. Needless to say, the inhabitants of Arnhem Land were never consulted about its visitation to their country.

Setzler's excavation of mortuary sites, conducted in various parts of Arnhem Land, is thoroughly documented in the official *Records* of the Arnhem Land Expedition and in his private papers. The latter are particularly incriminating, in that they candidly confess to the ruses he went to in avoiding the scrutiny of locals when he took the bones. Two young men, Jimmy Bungaroo and the other identified only by his mission name, Mickey, had been assigned to Setzler as archaeological assistants. His diary records that on 9 October 1948, accompanied by one of these men, he entered a large cave on Injalak where a skull and numerous long bones, all coated with red pigment, were visible. Setzler wrote: 'I paid no attention to these bones as long as the native was with me' (Setzler, 9 October 1948). On 28 October (the same day he filmed with Walker, although this appears to have been a separate incident), he writes of Bungaroo and Mickey:

> [d]uring the lunch period, while the two native boys were asleep, I gathered the two skeletons which had been placed in crevices outside the caves. These were disarticulated ... and only skull and long bones. One had been painted with red ochre. These I carried down to the camp in burlap sacks and later packed in ammunition boxes. (Setzler, 28 October 1948)

Setzler's surreptitiousness leaves little doubt of his awareness that even by his own standards, this was theft. Yet he took something else when he stole the bones: a photograph of Bungaroo, soundly sleeping on his bed of rock. He is dressed only in a *naga* (loin cloth) and near the lower margin of the photo are his pipe and tobacco tin, a reminder that those who laboured for the expedition were mostly paid in tobacco. Why Setzler photographed the sleeping youth is not explained. He did caption it minimally: 'Jimmy (Bungaroo), my native assistant from Goulburn Island asleep on a rock.' Perhaps he took it to illustrate the tale of how he outwitted the natives when the bones were won. Or perhaps this furtive gesture was a mere extension of the trophy hunting exhibited in his broader quest. Because the sleeper could not authorise or pose or in any way control his portrayal, the photograph displays with rare lucidity the power relations that existed between him and the man with the camera. Yet as much as it objectifies

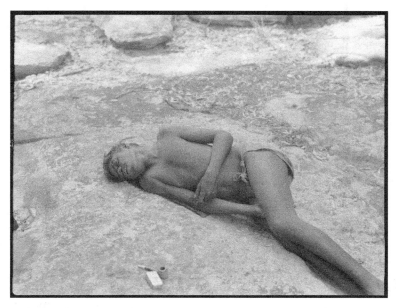

Figure 4. Jimmy Bungaroo asleep, 1948. Photograph by Frank M. Setzler. By permission of National Anthropological Archives, Smithsonian Institution (Photo Lot 36 oenpelli_142).

the youth, the eroticism of the image is unavoidable. Bungaroo is beautiful in his tranquillity. The sense of desire that saturates this image can (and perhaps should) be read sexually. Yet it also speaks of the likelihood that a hoarder of the dead will see in the living the dead of the future. That is why this photograph is such a powerful expression of the bone taker's gaze.

The battle to get the bones back is too long a story to tell here in its entirety. It took about a decade from the late 1990s, involved lobbying at an intergovernmental level, and provoked enormous hostility from certain factions at the Smithsonian, particularly within the National Museum of Natural History, an institution deeply divided about the status of its large, transnational collection of human remains. What concerns me here is the perspective of people including Wamud who had to grapple not only with the hurt caused by the theft of the bones but with the considerable challenges posed by their return.

Figure 5. Equipment and collections, including the boxed human remains, leaving Gubalanya at the end of the American-Australian Scientific Expedition to Arnhem Land, 1948. Photograph by Frank M. Setzler. By permission of the National Anthropological Archives, Smithsonian Institution (Photo lot 36 Oenpelli_160).

The repatriation of the Arnhem Land remains occurred in two instalments. Initially the Smithsonian would only release two thirds of the bones collected in 1948, citing an original agreement, brokered by Mountford, that all the collections amassed by the expedition were to be split at the ratio of two thirds to one third, with Australia, as the host country, the major recipient. Since the entire stock of human bones had been exported to Washington at the end of the expedition, the Smithsonian felt justified in releasing the two thirds in 2009—an outcome that was by no means acceptable to the traditional owners.

Later that year the Arnhem Land Expedition was the subject of a symposium at the National Museum of Australia.[3] The Smithsonian's insistence that it would keep in perpetuity the final third of the human remains was the subject of heated debate, and Thomas Amagula, a delegate from Groote Eylandt, issued a feisty press release comparing the Smithsonian's retention of his ancestors' bones with the US government's efforts to repatriate its own servicemen who had died abroad (Amagula). Kim Beazley, then Ambassador-designate to Washington and a speaker at the symposium, issued a statement that he would personally lobby for the return of the bones when he reached the US. Negative press coverage resulting from the symposium seems to have shamed the Smithsonian into voluntarily releasing the remaining third of the bones. This second repatriation took place in mid-2010. With Adis Hondo and camera, I travelled to Washington to document the homeward journey. Representing the localities from which bones were stolen were Joe Gumbula, a senior Yolngu man who spent his early years on the island of Milingimbi, Thomas Amagula, and Victor Gumurdul from Gunbalanya. To the bones they had a double duty: to both bring and sing them home.

The National Museum of Natural History and other major galleries and museums that comprise the Smithsonian Institution are situated along The Mall: a sacred axis in the religion that is America. The Mall commences at the foot of the Capitol Building and ends at the George Washington Monument. It was *not* the location offered to the visitors from Arnhem Land to ceremonially receive their ancestors' remains. That happened in the town of Suitland, a drab little satellite of the capital where the real estate is cheap and where the Smithsonian, when it outgrew its storage space in the downtown area, built climate-controlled facilities to warehouse collections. The premises are heavily guarded and encircled by an unscaleable fence. Suitland is 'A CARING COMMUNITY' according to the road sign, but when, a few years ago, I was going there daily to research the Arnhem Land collections, colleagues warned me that as a white visitor to a neighbourhood that is desperately poor and 93 per cent African-American, I should watch my step.

The gated world of the Smithsonian compound, barricaded against the racial-political reality of its own doorstep, was in a funny way an apposite locale for the ceremony that heralded the dispatch of the bones, which curators had packed in cartons that I came to think of as the 'cardboard coffins'. Ceremonies in the museum world are generally to do with openings, bequests or major acquisitions. But here we had come to celebrate, if that's quite the word, a less publicised aspect of the museum business: the departure of objects from collections—known as 'de-accessioning' in the curatorial jargon. I say 'objects' with a sense of reservation. Objects, after all, are *things*, whereas coffins—even if they are only cardboard boxes—are containers of *people*.

There were four cardboard coffins, each draped by an Aboriginal flag. Wheeled from the Smithsonian premises on a trolley, they led a stream of diplomats, officials and hangers-on, gathered to honour their departure. Kim Beazley, now ambassador, and Native American employees of the Smithsonian, who warmly supported the Arnhem Land visitors, had spoken movingly. Now outside, as clapsticks pounded like a heartbeat, Joe Gumbula ignited a handful of leaves, harvested from a Washington gum tree, and the smoking phase of the ceremony began. There were 50 or so of us milling around in the sunshine. We were a mourning party, for this was, after all, the beginning of a very long funeral. Gumbula, as the elder figure among the three Arnhem Landers, had performed from a repertoire of mourning songs, traditional to Northeast Arnhem Land. The light was exquisitely limpid on this northern summer day. A pair of armoured helicopters flew overhead, like hornets intent upon their destination. That was the President and his ubiquitous decoy, someone explained.

The bones travelled with us from Washington to Los Angeles and from Sydney to Darwin. Adis and I flew on to Gunbalanya with Victor and it was at that point that our work with Wamud truly began.

*

Figure 6. Draped in an Aboriginal flag, the boxes of bones leave the Smithsonian Institution. Digital photograph by Adis Hondo.

Beliefs about reincarnation and the origin of spirits among the Gagudju (Kakadu) speaking people of Gunbalanya were described by the anthropologist Baldwin Spencer, based on his fieldwork of 1912. As he explained it, the people of the present are reincarnations of the first people who inhabited the country. The spirit, known as the Yalmaru, has a cyclical existence, treading back and forth between the worlds of the living and the dead. When a person dies and the mourning ceremonies take place, the Yalmaru watches over the bones in a role of guardianship. With the passage of time, the Yalmaru divides into two parts. One half remains the original Yalmaru; the other is a discrete though inextricably connected entity, named Iwaiyu. 'The two are distinct and have somewhat the same relationship to one another as a man and his shadow, which, in the native mind, are very intimately associated,' according to Spencer (270). After a long period of watching over the bones, the time to reincarnate arrives. Iwaiyu assumes the form of a frog who, with Yalmaru's aid, finds and enters a woman. Iwaiyu develops into a baby and in this way the spirit returns to the domain of the living while retaining a presence in the world of the dead.

Contemporary beliefs in Gunbalanya, as Wamud explained them, do not exactly equate with Spencer's account. Much has changed in the past century and Wamud's family is among the few who can claim lineage with the Gagudju whose estates lie to the west of the town. Christian influence has its effect here, but in the syncretic cultures of Arnhem Land, the Bible can be treated as a belief system that sits alongside, rather than overriding, the Aboriginal traditions. So despite the many changes dating from the twentieth century, there are significant areas of continuity between the beliefs articulated by Spencer and those of today. Preeminent among them—and this applies to much of Aboriginal Australia—is the ongoing relationship between a person's spirit and their physical remains.

When seated near the cave with the 'man in his country' whom he wanted us to film, Wamud explained to camera the importance of being put to rest within one's territory after death. He explained that in earlier times, the body of the deceased went through various stages of processing. First, it was put on a platform constructed in a tree and covered in skins or, more recently, blankets. Later the bones were transferred to an elaborately painted hollow-log coffin, planted upright in the ground. This was part of the funeral rite known as Lorrkkon. Much later, the bones were removed and installed in caves or crevices. Wamud explained:

> They would get all bones, they would paint them in either yellow or red ochre, and then they would take them up onto the hill and put them in cave and they would say to them ... in language: 'I'll leave you here,' and talk to them, and then they would tell them, 'Look I'm following you, I'm coming too, so wait for me.' And they would put them to sleep then in their own country, in their own territory. Doing all that first, putting someone in the cave, doing that, that was the main traditional owner in there, or elderly person, who was the last one, and then [the] next one took over, maybe his son, and then it went on ... [We do it]

Figure 7. Argaluk, the sacred hill from which many of the bones were stolen in 1948. Digital photograph by Martin Thomas.

> not because we love you and we leave you here but because it's your country, we'll come back to you and we'll call out for help if you can help us ...
>
> Having the bones here it means that you've got the man still staying in his own territory ... I can call to him any time because I know he's here. No other questions ... I know he's here, I go to that cave and call out for help. Like somebody might be in danger. That's when you really need someone. So the traditional owner would help and in spirit they would still help.

What I have come to realise, in talking about these issues, and in being taken to sites where remains were stolen, is that the dead never become objects or object-like. This is a fundamental contrast to the Judeo-Christian tradition where death marks a rupture between flesh and spirit; where the soul goes elsewhere. In west Arnhem Land, I have found that the precise relationship between spirit and bones is never easily expressed, perhaps because of the limits of my questioning or understanding, because of the linguistic differences, or because there is a range of beliefs about such matters. Most likely it is a combination of them all. One reason why the behaviour of spirits is not readily pinned down is because spirits have agency. Like living people, they can be kindly or malicious; friendly, dangerous or merely indifferent; healers or spreaders of disease. Accidents or mishaps are often attributed to spirits wreaking mischief. So care is required, even in what you say about them.

As I understand it, spirits live in proximity to bones, though they are not embodied within them. Spirits and bones are intimately related. Although one is immaterial and the other material, neither will perish. The living and the dead are co-existent. In this place-based cosmology, it is only natural that a person's bones should be laid to rest in their ancestral estate. For Bininj, the actions of Frank Setzler were a brazen theft, and it is not really surprising that when the stolen bones were buried on the edge of Gunbalanya in 2011, Wamud lectured severely in English to all assembled (not ignoring the camera from a TV news crew): 'Stealing people's bones for study is no bloody good.' But to confine the transgression to the act of theft is to understate the gravity and complexity of the problem. Theft is a crime against property, whereas this was a crime against people. The removal of bones is closer to kidnap from this point of view. Taking the bones presented Bininj and Aboriginal people from other parts of Arnhem Land with the terrifying possibility that spirits had been wrenched from their country and taken abroad on what Wamud called a 'long walkabout'. Now the Bininj must deal with the bewilderment and possible anger of the spirits upon their return to Australia.

*

It took twelve months for the Bininj to decide how to deal with the repatriated bones. During that period they were locked in a shed attached to premises occupied by Bill Ivory, Gunbalanya's Government Business Manager. In addition to the two repatriations from Washington, remains returned by other institutions, among them the Museum of Victoria, had been added to the pile of boxes. A few weeks earlier, Steve Webb, a physical anthropologist from Bond University, visited the town to sort the bones. The Bininj avoided handling or even looking at

Figure 8. Isaiah Nagurrgurrba harvests paperbark for wrapping the bones. Digital photograph by Martin Thomas.

them for as long as possible. They seemed happy to delegate the technical work to Balanda, and Steve's contribution signalled a significant transition in their recognition as human subjects. The bones had been stolen in the service of physical anthropology, so perhaps it was appropriate that the specialised knowledge of that discipline should be brought to the service of the bones themselves, now that their long walkabout was ending. At the request of Wamud and others, Steve was asked to identify the sex of the remains and, insofar as he could, to bundle together the bones of individual people. Wrapped in paper, no longer specimens but now individual men, women or children, a Texta colour arrow pointed in the direction of the head (though in some cases the skull was missing). Steve then returned the bundles to their cardboard boxes. They were in this state on 19 July 2011 when, with a plentiful supply of paperbark sheets, torn from trees the day before, preparations for burial began.

Reuben Brown, a PhD student who is studying sound recordings made on the Arnhem Land Expedition, fetched Wamud from his house, pushing him through town in his wheelchair. Wamud's son Alfred Nayinggul and other senior men and women were waiting, as were Suze and Adis with mic and camera, and Glenn Campbell, a photographer representing the Fairfax media group, whose work that day would win him a Walkley Award. A fair amount of discussion had occurred about where the wrapping should take place. Assuming that privacy would be required for proceedings so sensitive, we had used sheets and tarps to screen off a section of enclosed verandah that was part of Bill's living quarters. As with many things that were apparently pre-decided, Wamud reversed his

Figure 9. Isaiah Nagurrgurrba (left) and Victor Gumurrdul dig a grave for the repatriated bones while Reuben Brown and Alfred Nayinggul look on. Digital photograph by Martin Thomas.

position as events unfolded. The boxes, he declared, would be opened outside in the shelter of a carport. Nothing would be hidden. Only the previous day, I had asked Wamud what, if any, of the preparations could be photographed. He said that we should start filming just before the bones were about to be wrapped.

Overnight, however, he had decided that the proceedings should be documented more thoroughly. I heard later that Glenn Campbell's photographs in *The Age*, in which the bone handling was depicted in all its stark intimacy, were considered culturally inappropriate by gatekeepers in the metropole. In fact all of this was intentionally made open to the public gaze. A blue tarpaulin, laid down on the ground, formed the photographic backdrop. The large pieces of paperbark were brought forth and pulled apart into thinner layers, establishing the foundation for the work that followed.

When the shed was finally unlocked we gazed at the stack of boxes. As documenter of the repatriation, I don't know that I'd ever been much of an omniscient observer, but from this time I became ever more a participant in the proceedings. Alfred, Reuben and I began to remove the boxes. As tends to happen in Bininj-way if you hang around long enough, you become part of the action—at least to some degree. We carried the boxes, though significantly we never handled the bones. I'm sure Alfred would have preferred that we'd done so. He had said to Bill the previous week how nervous he was lest he should be asked to touch them. He was fearful at getting sick or being otherwise molested by the spirits. As Alfred dithered before a box of bones in the carport, Wamud barked at him from the security of his wheelchair and told him in effect to get his act together: to get those bones out of the boxes and prepare them for burial.

At a recent conference, I showed some of this film. Of course I had to edit severely, removing from sight the figure who was, and who in time again will be, at centre stage. In the doctored footage, even without his visible presence, his power over the scene is palpable. Or perhaps it is made more obvious *because* of his invisibility. For what is a taboo if not a form of emphasis? In editing the sequence, Adis and I manipulated not only the visual but the *audible* record, for Wamud's chief role in these proceedings was not only to command the living as they went about their labour of tending to the dead—this he did with authoritarian severity—but to talk to the spirits, to calm them, assuage them, to tell them where they were. I believe that the necessity of doing this had long preoccupied him, and it explains why he alone, among the few veteran old people of the community, was qualified to perform the role he did. Lorrkkon, the hollow log ceremony, is now defunct, but elements of it are evident in the way the bones were ochred and wrapped. But most of all it was Wamud's linguistic heritage that made him uniquely qualified to perform the work required on that day. Gagadju, Urningangk, Erre and Mengerrdji were the languages native to Wamud's ancestral estates around the East Alligator River, and the ones in his estimation likely to be most familiar to the spirits. Competency in those languages has fairly much died with him, although I suspect that even his was little more than a remnant knowledge. Kinwinjku, which arrived in Gunbalanya with migration from the east has for the most part replaced these tongues, as has Kriol and—to some extent—English.

The days after the burial he did what he could to translate what it was he said to the spirits, but pain and failing strength prevented much progress. So I must add this failure to my unhappy list of uncompleted tasks. But the gist of it he paraphrased. As each box was opened, he began his incantations. He told the spirits his name. He told them of the major landmarks: Argaluk (towards which their head would lie) and Injalak (to which their feet would be pointed). He told them they had come home and would be laid to rest, not in the caves from which they were stolen, but deep in the ground where no one could take them again.

It lasted the greater part of the day, the preparation of the bones. Wamud maintained his intercourse with the spirits. The men worked on men's remains; women worked on women and children. Red ochre was ground and mixed with water, and not merely applied to the bones, but massaged into them. Wamud said to me as he presided over this scene of industry, immeasurably satisfied, that they were 'dressing' them. Painted with the pigment of the earth, as in life they had been painted for dance, it was a final kindness from living hand. Fibulas and tibias were quickly coated with a hand plunged into the red solution, rich as blood. The smooth domes of crania were coated, quickly drying to a powdery rust; eye-sockets were daubed with dripping fingers. The discovery of a stray tooth had Donna Nadjammerek, one of the embalmers, testing socket after socket until at last she restored it to its owner.

The bundles of bones we had in a way been ready for, but the plastic bags defied all expectation. Inside them were the tailings from the archaeologist's sieve: humanity reduced to splinters and fragments. Big bags bulged with larger pieces, like coarse gravel. Tiny bags, no bigger than drug deals, held quotas of grit. No hope of painting this stuff. After what seemed interminable emptying onto their cushiony layer of paperbark, they were quickly bundled and tied, and the bags, like all the boxes and wrappings tainted by their contact with the dead, were put aside to be taken to the grave site next day. Ashes to ashes. Dust to dust.

*

The ceremony that ended the long walkabout began with a procession through the town. To my great honour, Wamud asked that I push him in his wheelchair, and as we set off with Bill, Alfred, Donna and Wamud's daughter Connie, and with politicians, dignitaries and others in tow, he barked at me constantly to walk slowly, more slowly, slower still. With singers and didjeridoo beside us, guiding the spirits in a current of song that would eventually coax them into the confined space of the grave, we maintained a funereal pace. The boxes of bones were following in a cortege of four-wheel drives. People watching from their houses fell in behind the vehicles as the march proceeded. All were invited: Bininj, Balanda, visiting workers and tourists shopping at the art centre. At the grave site, Wamud spoke at length, in various languages, to the assembled audience and the spirits. To the latter he explained that they had been divided into men and women, Bininj and Daluk, just the same as if they were camping.

The bundles were buried in deep trenches and filled with earth. As the crowd dispersed, and headed off for a barbecue laid on in celebration, the boxes and wrappings that had held the bones were torched. A flash of orange, and the cardboard coffins were no more. Their destruction, as much the burial itself, confirmed that the status of the bones as things in a museum had irrevocably ended.

I was there a month later, without the film crew. It was the last time I saw Wamud, who was largely confined to the horrific squalor of his bed. I recorded an interview in which he said that he was immensely satisfied with the job we had done. By this he meant the film, of which I had shown him rushes, as well as the greater project of restoring the spirits to their country. I asked whether the film was intended as a guide to those who might be dealing with repatriations in the future. He said that was among the reasons. But foremost, I think, was the desire to say something about his life and memories, and to perpetuate his lifelong interest, dating from his childhood on the mission, in communicating across cultures. Arnhem Land is rich in many things, poverty among them. He wanted to open something of his world to you and me.

He is buried now in his cherished homeland, Mikkinj Valley, southeast of Cahills Crossing. He often spoke of this place, and had it been possible, he would have died there. But when, in July 2012, they found room for him in the Gunbalanya morgue, he was brought back to town and laid to rest at last. There was the big

Figure 10. Charred remnants of the cardboard coffins. Digital photograph by Martin Thomas.

church service in Gunbalanya. And then—to use *your* words Wamud—they took you to Mikkinj *because it's your country.*

Hundreds made the journey to camp with him as his body lay resting in a shelter made of boughs. They sang for him night and day, and each and every mourner, black and white, went to sit with him, and in their own way, in their own time, in their own language, each said goodbye. As they buried him, small whirlwinds were seen dancing through the campground and immediately recognised as his spirit.

At night during these proceedings young men patrolled the border of the camp and periodically called out to him in Kunwinjku, telling him to go back. 'Go back! Go back!' they cried. 'Go back!' Why go back? Wasn't he already back? All finished up and back in his country? Yes indeed. But they meant back to the spirit world, back to the place from which he started, and to which, we might suppose, he has now returned. For the moment we must leave him, at home in his country, and avoid for a time the temptation to recall him. This beautiful country. *Because it's your country.* Troubled and majestic, battered and beloved; this is his country, the greatest teacher of them all.

Acknowledgement

This was the winner of the 2013 Calibre Prize for an Outstanding Essay, awarded by Australian Book Review. *Martin Thomas, who is an Australian National University historian, thanks the editor of* ABR *for allowing republication of the now peer-reviewed essay. Thanks also to the community of Gunbalanya and to the Australian Research Council for financial support.*

Notes

[1] Statement recorded by the author on 1 September 2006 at Kungarrewarlk outstation, near Gunbalanya (Oenpelli) NT. Translation by Murray Garde, 22 September 2006.
[2] See Thomas, 2010; Thomas and Neale, 2011; and May, 2010.
[3] http://www.nma.gov.au/history/research/conferences_and_seminars/barks_birds_billabongs/home. Accessed 20 December 2012.

References

Amagula, Thomas. 'First forgotten Australians call for return of human remains.' Media statement dated 19 November 2009. 20 January 2013. <http://www.anindilyakwa.com.au/periodicals-and-annual-reports>

Australian Bureau of Statistics. *Experimental Life Tables for Aboriginal and Torres Strait Islander Australians.* 2009. ABS Publication 3302.0.55.003, 7. 12 December 2012 http://www.abs.gov.au/AUSSTATS/abs@.nsf/DetailsPage/3302.0.55.0032005%E2%80%932007?OpenDocument

May, Sally K. *Collecting Cultures: Myth, Politics, and Collaboration in the 1948 Arnhem Land Expedition.* Lanham, MD: AltaMira, 2010.

Spencer, Baldwin. *Native Tribes of the Northern Territory of Australia*. London: Macmillan, 1914.

Setzler, Frank M. Diaries 1948, Box 14, Frank Maryl Setzler Papers 1927–1960, National Anthropological Archives, Suitland MD.

Thomas, Martin. 'A Short History of the 1948 Arnhem Land Expedition,' *Aboriginal History* 34 (2010): 143–70.

Thomas, Martin and Margo Neale, eds. *Exploring the Legacy of the 1948 Arnhem Land Expedition*. Canberra: ANU E Press, 2011.

Walker, Howell. Cinematographer. *Aboriginal Australia* (lecture film). 1950. National Geographic Society, Washington DC.

'from Organic Acts': Tsamorita, Rosaries, and the Poem of My Grandma's Life

Craig Santos Perez

This creative nonfiction essay shares the process of writing my poem, 'from organic acts,' which was published in my second book of poems, *from unincorporated territory [saina]*. The poem is about my grandma's life, from her early childhood growing up on Guam during World War II to the point in her seventies when she migrated to California. In the essay I discuss the process of interviewing my grandma, writing the poem, and incorporating sources related to Chamorro orature, Catholic rosaries, US citizenship, and intergenerational storytelling.

'Gi na'ån i tata, i lahi-na, yan i espiritu sånto, åmen' (Rosario's Rosaries), my grandma recites, clutching her rosary. I sit next to her on the couch and whisper her name, Milan Martinez Portusach Santos Reyes, as she prays to the makeshift altar of votive candles and crucifixes in her small apartment in Fairfield, California. 'Umatuna i Tata, I lahi-na yan i espiritua sånto. Taiguihi i tutuhom-na yan pågo, yan siempre yan i minaihinekkok na ha'åne, åmen', she continues. Her voice and the words of our native language form a canoe that carries me on the currents of memory back to our home island of Guåhan (Guam) in the western Pacific Ocean.

Spain colonized Guåhan in the seventeenth century, establishing the first Catholic mission in the Pacific. Catholic prayers and the Bible were translated into the indigenous Chamorro language. From these colonial translations, the 'techa' emerged: a prayer leader who directs the rosary, novena, and other devotions. The word 'techa' comes from the verb 'tucha', which means 'to recite aloud' or 'to lead in prayer'. The techa chants the rosary in a steady, nasal monotone; at certain moments, the listeners respond (Tolentino).

My grandma isn't a professional techa, but she does recite the rosary every day. I watch her press the rosary beads between her fingers, like waves pressing stones. Her mom taught her how to make and recite the rosary. 'I really miss my mom', Grandma says. 'She's very devoted to the blessed mother / every evening at six we had to come home / and say our rosary / in front of the Madonna'.

After the 1898 war between Spain and America, Guåhan became a possession of the United States. The colonial administration passed English-only laws, established an American education system, and introduced American music, literature, and media, all of which marginalised Chamorro language. Additionally, Catholic church services began to be performed in English. 'We went to school and we learned / how to speak English', Grandma says. 'We learn all the grammars but at home / we say rosary in Chamorro'.

The rhythm and intonation of the Chamorro rosary is related to an ancestral form of Chamorro oral poetry known as *tsamorita*, an improvisational and communal song poem. Here's how the form works: a poet sings a four-line verse (each line containing eight syllables), with the second and fourth line rhyming. Then a second poet repeats the last two lines and adds a new couplet, rhyming their last line with the previous poet's last line. The poets continue throwing verses back and forth. This form was employed to narrate ancestral history, cultural values, and religious beliefs, as well as to perform social entertainment, political debate, and romantic courting. The tsamorita could be heard during communal activities, such as fishing, planting, harvesting, canoeing, and weaving. The tsamorita tradition changed under Spanish rule, when Christian themes dominated and Chamorro history and culture were repressed. Adapting to these changes, the tsamorita voice found new expression in the voice of the techa (Flores *Guampedia*). 'Manhongge yo' as Yu'us Tåta ni todu ha' hana'sina, na hana'huyong i langet yan i tano'', Grandma continues.

In 1941 the military of Japan invaded and occupied Guåhan for three years. 'We are at home', she says. 'And then we heard siren and people / telling us that the Japanese are invading / so we ran up the hill and we ran'. Her body tightens as she struggles to tell this story. 'Almost every night during the war neighbours come to our house / In the basement / my mother // would translate the radio / that someone brings to the house / because they had it buried / somewhere during the day'. Her alarm suddenly rings because it's time for her to take her pills. She opens her pill organiser, and I give her a glass of water. Then she whispers, 'My mom always makes a huge pot of soup / for everyone // But we had to be / very quiet / because you never know / who's outside / in the dark'. As the war intensified, fewer neighbours attended the night gatherings. Some neighbours she never saw again.

When I attended both public and Catholic schools on Guåhan, I never learned about Chamoru culture, history, or language. Thus, my grandma's stories became my history. Her voice carried our memories, and I learned to read those memories for their cultural knowledge and values. Even if I didn't understand all the words, I could listen to her recite the rosary and immerse myself in the waves of our now endangered language. I became a poet after years of listening to my grandma's stories. I write, predominantly in English, the stories my grandma passed down to me. Poetry houses her stories as written memory, as memorial. Poetry throws her verses towards new audiences, listeners, and spaces. Poetry weaves our voices together.

During the past year, my grandma has suffered from severe joint pain, as well as weight and memory loss. I wrote a poem to help me process the visible signs of her aging. Two images haunt: 'her legs barely as thick / as her arms' and 'her veins resemble / wings / caught / in nets'. Because I have always been caught in the strength of her voice, writing these images was the first time I felt seized by the organic nature of her body. The poem also juxtaposes religious images of the cross, the rosary, and the Virgin Mary. The poem ends: 'crown of roses—rosary—rotary—rot'.

As her health continued to deteriorate, I didn't want death to take her body. I didn't want all the stories she has yet to share with me to become lost. I didn't want her body frozen at 'rot'. If I revised the poem, could I revise her aging? If I told her story in a new way, could her story never end?

My first act of revision was to record her voice onto the poem. With her permission, I began writing her words as she spoke, breaking the line when she paused, breaking the page when she took a deep breath, breaking the words when her voice broke. Her words animate the poem, just as they animate this essay, and give her agency and presence. 'Every night we had novena / I think that's what really helped us during war / our devotion / to blessed mother', she says. As we moved through the stations of the cross, excerpts from the Chamorro rosary entered my poem.

Just as my grandma's voice inhabits her body in the poem, I wanted history to inhabit the religious relics and imagery. My second act of revision was to include a historical timeline in the poem that mapped how Catholicism became rooted in Guåhan. The timeline not only historicises the church and its artefacts, but it also acts as a kind of historical stations of the cross. Grandma says, 'Every night we had novena / I think that's what really helped us during war / our devotion / to blessed mother. I'm still carrying / what I learned from my mother // I'm still carrying it / for the sake of us'.

There's another story my grandma carries with her. Once, a giant fish began eating the centre of our island. Men from the villages hunted, with spears, in their canoes, but the giant fish easily hid from them. That's when the Virgin Mary came ashore. She wove her hair into a net, caught the fish, and saved the island. There's an older version of the story. Once, a giant fish began eating the centre of our island. Men from the villages hunted, with spears, in their canoes, but the giant fish easily hid from them. Young women from the villages wove their hair into a net. With beautiful voices they sang the tsamorita and lured the fish to the surface. They captured the fish and saved the island. My third act of revision was to collage the different versions of this ancestral story found in Mavis Warner Van Peenan's *Chamorro Legends on the Island of Guam* and Eve Gray's *Legends of Micronesia: Book Two*. This story creates a mythic and deeply ancestral space in the poem, while also speaking to how Catholicism changed not only how we tell stories, but our stories themselves.

My grandma was born in 1928. During that time, Guam was an unorganised, unincorporated US territory governed by the US Navy. A civilian government, administered by the Department of the Interior, was created through the 1950 Guam Organic Act, and US citizenship was granted to island residents. The Organic Act also lists the land and acreage that the US military would continue to control and occupy—nearly 30 percent of the 212-square-mile island. My third act of revision was to collage into my poem two sections of the Organic Act related to land takings. This includes 'real property', as well as 'tidelands' or 'submerged lands', lands permanently or periodically covered by tidal waters. Military fences become the teeth of the giant fish.

My last act of revision: I inscribe myself, as grandson, into the poem. I am present on the couch listening to her story; I help her stand and walk between couch and kitchen table; we eat dinner, drink tea. As our living, intimate presences entwine on the page, I feel the objective detachment of the narrator become the intimate attachment of a participant. I am part of her story because my survival is testament to her survival. Her life and voice will continue in my words; our lives and voices—even after we are both gone—will continue in the poem.

I titled the poem '*from* Organic Acts'. It is 50 pages long, divided into five excerpted sections, and dispersed throughout my second book of poems, *from unincorporated territory [saina]*. The seriality of the poem echoes the stations of the cross, or what I consider stations of poetic crossing. The excerpted form suggests that the story has not yet been fully told; thus, it has not yet ended. The origin and final destination word of the poem, along with the story of my grandma's life, remain as organic possibility.

In 'Some Notes on Organic Form', American poet Denise Levertov describes her poetic practice as a 'method of apperception' (of recognising what we perceive) based on 'an intuition of an order, a form beyond forms, in which forms partake, and of which man's creative works are analogies, resemblances, natural allegories' (Levertov, 68). For me, the poem is an allegory of crossing. The page is an ocean; words are islands; sentences are archipelagos. Form is never more than a navigation of content.

Levertov describes her poetic rhythms as 'strands of seaweed moving within a wave' (71). I hear the rhythms of my poems as waves crashing against the shore of attention, enlivening the tidelands between poem and reader. Levertov also likens the poet to a helicopter scout flying 'over the sea to watch for the schools of herring and direct the fishing fleet toward them' (72). In writing my poem, I merely felt like a paddler in an outrigger canoe, surrendering to the winds, currents, shadows, stars, dreams, distances, memories, and stories of my grandma's life.

After I completed the final, ocean-ready form of the poem, my family learned that my grandma was suffering from Alzheimer's. She moved in with my parents in Sunnyvale, California, so that they could care for her and she could attend a nearby Alzheimer's care facility during the weekdays. When my book was published, I brought the box of books over to my parents' house, and my auntie (my mom's younger sister), happened to be visiting my grandma. I opened the box, gave them all a copy of my book, and (to my embarrassment) they immediately began reading my new book. Suddenly my mom exclaimed to my grandma: 'Look, it's your name! You're famous now!'

We all sit in the living room. My mom and aunt begin taking turns reading '*from* Organic Acts' aloud to my grandma. Grandma listens, intently, to her story. When the first section ends, my auntie asks, 'What happens next?' They instinctively flip through the pages, skipping the other poems, until they reach the next excerpt of '*from* Organic Acts'. 'Here', my auntie says. 'It continues'. When they reach the part of the poem about my grandma's childhood, they stop reading the poem and begin to ask my grandma about their own grandmother, about what it was like during the occupation, about those nights in the basement with the radio. Apparently, my grandma had never told her own daughters these stories. After she answers their questions, they return to reciting the poem.

Since that day, my grandma has continued to lose her memory. She has her own room in my parents' house, and she keeps a copy of my book on her bedside table, next to a crucifix. My mom tells me that sometimes Grandma grows scared at night in her room by herself because she doesn't remember where she is or who she is living with. My mom reminds her: you live in California now, with me, Helen, your oldest daughter. To help comfort her, my mom will sometimes read my poem '*from* Organic Acts' to her. As my grandma hears her life being read back to her in her daughter's voice, her memory of her life slowly returns. 'I really miss my mom', Grandma once said. 'Her voice is like my voice // when I say rosary I think I can hear her voice / even here in California'.

A master poet once described performing the tsamorita as 'being able to sing both forwards and backwards' (*Kantan Chamorrita Revisited*, Flores 22). In the tidelands of the future and the past, poetry becomes an intimate space between memory and memorial, an intimate space between generations, and an intimate space between words and silence, voice and loss. I believe that poetry, in our darkest moments of forgetting, in the last years of my grandma's life, will help us navigate the organic acts of dying. Her voice will forever echo: 'Santa Maria, nanan Yu'us, tayayute ham ni i manisao pago yan i oran finatai-mami. åmen'.

References

Flores, Judy. *Guampedia*. 'Kantan Chamorrita', 2009. 17 May 2013. <http://guampedia.com/kantan-chamorrita-2/>.

Flores, Judy. 'Kantan Chamorrita Revisited in the New Millennium'. *Transnationalism and Modernity in the Music and Dance of Oceania: Essays in Honour of Barbara B. Smith*. Ed. H. Lawrence. Sydney: U of Sydney, 2001. 19–31.

Gray, Eve. *Legends of Micronesia: Book Two*. Honolulu: Trust Territory of the Pacific Islands, 1951.

Levertov, Denise. *New & Selected Essays*. 1965. New York: New Directions Publishing Corporation, 1992.

Rosario's Rosaries. 'Rosario's Rosaries: The Rosary Prayers in English and Chamorro', 2008. 7 Dec. 2014. <http://charoanderson.com/rp-engcham.html>

Santos Perez, Craig. *From Unincorporated Territory [saina]*. Richmond, CA: Omnidawn Publishing, 2010.
Tolentino, Dominica. *Guampedia*. 'Techa: Traditional Prayer Leader', 2009. 17 May 2013. <http://guampedia.com/techa-traditional-prayer-leader>
Van Peenan, Mavis Warner. *Chamorro Legends on the Island of Guam*. Mangilao, Guam: Micronesian Area Research Centre, 1974.

Index

Aboriginal human 129
Aino Pargas, epistolary dynamics of sisterhood *see* Helga Sitska and Aino Pargas
Amagula, Thomas 133
American-Australian Scientific Expedition to Arnhem Land 125, *125*
American education system 146
'analogue' era 114
ancestral story 147
Anna Dijkman-Tetteroo 57, 89; food drop 84; impressions of Australia 91; wartime quilt 79–81, *80, 81,* 83
anthropological study 125
Anti-Social Family, The (Barrett & McIntosh) 21
archaeologist 130
Argaluk Camp 128, *136*
Arnhem Land: Aboriginal people of 124; ceremonial knowledge in 125; Judeo-Christian tradition 136; map of *125*; official records of 130; pan-regional Kunapipi ceremony 126; syncretic cultures of 135
Arshad, Rowena 37n3
Arthur, Paul 3
Augé, Marc 118
Australian beach holidays *87, 88*
Australian immigration propaganda *90*
Australia quilt 57, 58, *60,* 84, 85
auto/biographical truth 2
autobiographical writing: case study of 101–6; categories of testimonials 99; definition 98, *99*; document humaine 99; traumatic memories in 98

Bachelard, Gaston 118
Baker, Mark Raphael 11
Banks, Olive 22
Barbour, John D. 15, 16
Barrett, Michele 21
Barthes, Roland 99
Beazley, Kim 133
Bechdel, Alison 12–16, 21
Bender, Barbara 66
Benjamin, Walter 110, 115, 118n4

'Bininj way' 124, 137
black feminists 21
bone handling 139
British National Life Stories collection 20
Brownie, Tiny 65
Brown, Reuben 138
Broyard, Bliss 8, 9, 11, 13–15
Bruss, Elizabeth 2
Buitenkampers 73–5
'Bulanj' 124
Buss, Helen 21
Butler, Anne 27

Cahill, Paddy 128
Cahills Crossing *127*
Calwell, Arthur 130
Campbell, Glenn 139
Catholic church 146
Catholicism 147
Catholic mission, in Pacific 145
Cecchetto, Vikki 4
censorship 100
Cerf, Vinton 110
Chamorro language 146
Chamorro Legends on the Island of Guam (Van Peenan, Mavis Warner) 147
Chamorro oral poetry 146
Chamoru culture 146
Chapman, Yvonne 57
Chester, Gail 27
childhood quilt 57, 58, *58*; de Certeau's philosophy 66–8; Frances Larder 62, *62*; Gerada Baremans 62, *63, 64,* 65; Ineke McIntosh-Eichholtz 61, 62, *63, 64,* 65; Johanna Binkhorst 65, *65, 66, 68*; Norberg-Schulz's philosophy 67, 68; sense of place 67, 68; Vicky van der Ley *67,* 68
Cixous, Helene 114, 115, 118
Claudel, Paul 115
Cockburn, Cynthia 33
colonial translations 145
Connor, Steve 28
Cormier, Ken 29

INDEX

Cosslett, Tess 21
cultural heritage, definition of 61
cultural-historical contexts 3
cultural memory 117
'cultural' mythology 100
cultural transmission 114
Czcibor-Piotrowski, Andrzej 4, 5, 98, 100; childhood memories 103; chronological plot development, lack of 101; Communist Poland 101; historical description 101–3; in Polish military 102; time frame of 101–3; translator 103; traumatic experiences 103; artime experiences 103; WWII deportees 107

'dark ages' 117
data preservation 117
data wealth 117
'death name' 124
Diecker, Ann 57
digital age 118
digital cameras 110
digital era 111
digital technologies 109, 111, 117
digital world 114
Double Oblivion of the Ourang-Outang 114
Dreams from My Father (Obama) 11
dubious cross-cultural honour 112
Dutch emigration propaganda *90*
Dutch Odyssey Quilts 4

Eakin, Paul 99
East Alligator River 127
'egodocuments' 99 *see also* autobiographical writing
Eksteins, Modris 9, 11, 16, 17
emotional memory 44
epistolary dynamics of sisterhood *see* Helga Sitska and Aino Pargas
'erasing external memories' 117
Estonia: 'Great Escape to the West' 53n3; life story collection in 53n1 (*see also* Helga Sitska and Aino Pargas)
Ethics of Life Writing, The (Barbour) 15
Evans, Paul 113–14
'eyewitness testimony' 100

Facebook/social media 117
face-to-face contact 3
Faithful Scribe, The (Mufti) 8
'Families and Children' 29
family memoirs 7, 17n1; academic historians 12; Barbour, John D. 15, 16; Bechdel, Alison 12–15; death of parents 8–9; Eksteins, Modris 9, 11, 16, 17; elements of 8; examples of 11; family photographs 11–12; Flem, Lydia 8, 12, 13, 15; Jewish families 11; *Lost, The* (Mendelsohn) 9–11; Mendelsohn, Daniel 9–13, 17; Miller, Nancy K. 8, 9, 11, 14; *One Drop* (Broyard) 8; 'positivism' 12; *Reading Autobiography* (Smith & Watson) 7; Rosenstone, Robert A. 8, 11, 13; Spiegelman, Art 11, 13, 14, 16, 17; Stille, Alexander 9–16; *What They Saved* (Miller) 8
Family of Man, The 23
family stories 20; British National Life Stories collection 20; Jones, Barbara 25–6; Taylor, Barbara 22–5; transformation of 21
Family Ties: Reframing Memory Exhibition 29, 31, 35
Female Eunuch, The (Greer) 27
Fiftieth Gate, The (Baker) 11
Final Reminder, The: How I Emptied My Parents House (Flem) 8, 11
First Nations peoples 114
Flem, Lydia 8, 12, 13, 15
food drops, wartime quilt 82, 83, *84–6*
Force of Things, The (Stille) 11
Frances Larder 57–8, 60; childhood quilt 62, *62*; food drop 85; life in the bush *89*; wartime quilt *69, 70*, 71
'from Organic Acts' 148, 149
'from unincorporated territory' 148
Fun Home (Bechdel) 12, 13
Funnell, Anthony 110

Gagudju (Kakadu) speaking people 135
Garde, Murray 125
Gerada Baremans: quilting piece *64*, 65; visual diary 62, *63*; wartime quilt 78, *79*, 81, *82*
Gerber, David 44
German airstrikes 102
Germany 102
Gluck, Sherna 26, 28
Greer, Germaine 27
Guam Organic Act 148
Gumbula, Joe 133
Gunbalanya 124, 125

Haley, Alex 10
'handed down' 114
Hanna, Martha 44
Hearn, Kirsten 27, 33
Helga Sitska and Aino Pargas 41, 53n2; bundle of letters 43–4; correspondence 41–4; dynamics of sisterly affection 42; emotional memory 44; epistolary exchanges 42, 44; intimate family relationships 52; preservation 43, 44; record of a lifetime 50–1; reliance on shared memories 47; sharing of photographs 49; sisterly roles 48–9; strategies of intimacy 46–8; writing in code 44–6
Henkel, Linda 118n1
Hirsch, Marianne 10, 21
Hondo, Adis 127, *134*

INDEX

Hoskins, Janet 115
Houghton, Suze 127
Hughes, Ed 33, 36
human bones 133

Ineke McIntosh-Eichholtz 57; quilting piece *64, 65*; visual diary *61, 62, 63*
Internet Archive's global Wayback Machine 116
Isaiah Nagurrgurrba *138*
'It Flies to the Hive' 48

Japan military 146
Jimmy Bungaroo *131*
Johanna Binkhorst 57, *93*; Australia quilt 85; food drop *86*; quilting pieces 65, *65, 66, 68,* 91; wartime quilt *68,* 69, 70, *71, 78, 82, 83*
Johnson, Rebecca 27
Jolly, Margaretta 2, 5
Jones, Barbara 25–6, 29
Judeo-Christian tradition 136
juxtaposes religious images 147

Kakadu National Park 127
Kirss, Tiina 49
Kitzinger, Sheila 27
Kodi stories 115
Kodjok 124
Kroll, Una 27
Kunwinjku speaker 125
Kurvet-Käosaar, Leena 4, 6n2

Lacy, Suzanne 27, 28
Lane, Cathy 27
Last Asylum, The (Taylor) 22
Legends of Micronesia 147
Lejeune, Philippe 2, 98
Leskelä-Kärki, Maarit 48, 52
Levertov, Denise 148
Lewis, Helen 117
Liddington, Jill 28
life writing 1, 2, 98
Liiv, Juhan 48
Lockwood, Annea 27
Lost, The (Mendelsohn) 9–11
Lury, Celia 21
Lynch, Kevin 67

MacColl, Ewan 29
MacCrum, Mukami 27
Marchant, Alison 27
material memory: data preservation 117; digital extension of 109; digital storage 117; letter writers feel 113; open-access principles 116; 'par avion' envelopes 112; Pavel's letters 112, *113*
'Matka' 104
Maus (Spiegelman) 11, 13

Mayer-Schönberger, Viktor 117
McIntosh, Mary 21
memory 111, 114, 118n7
Mendelsohn, Daniel 9–13, 17
Mertelsman, Olaf 53n4
'method of apperception' 148
Milan Martinez Portusach Santos Reyes 145
Miller, Nancy K. 8, 9, 11, 14, 16–17, 21, 52
Mitchell, Juliet 27
'Monople Magnums' 112
Morrison, Toni 44
Mother/Daughter Plot, The (Hirsch) 21
Mountford, Charles P. 130
Mufti, Shahan 8, 11, 16
Murray, Jenni 27

National Geographic Magazine 130
National Geographic Society film 129
National Library of Australia's Pandora project 116
National Museum of Natural History 131, 133
neurological systems 109
Norris, Jill 28

Oakley, Ann 26
Obama, Barack 11
Odyssey Quilting Project 61
Odyssey Quilts 57 *see also* Australia quilt; childhood quilt; wartime quilt; cultural heritage, notion of 61; remarkable achievement of 94
Oenpelli Mission 126
Olijnyk, Petro 111
Oliveros, Pauline 28
One Drop (Broyard) 8, 13–14
open-access principles 116
Opowieść w szesnastu snach (Czcibor-Piotrowski, Andrzej) 103
oral history *see also* autobiographical writing; family stories: and sound art 26–36; of women's liberation movement 21
'Oral History of British Science' 20
'orphaned memories' 110

pan-regional Kunapipi ceremony 126
'*par avion*' envelopes 112
Parker, Charles 29
Patai, Daphne 26, 28
Pavel's letters 112, *113*
Perez, Santos 3
personal trauma 101
Peters, Nonja 4
photographic portraits 111
photograph of Bungaroo 131
photographs 110
poetic rhythms 148
Poëzie booklet 61

153

INDEX

Popkin, Jeremy D. 4
Portelli, Alessandro 99
positivist scholarship 12
postmodernist scepticism 13
posttraumatic stress disorder (PTSD) 101
post-war extravaganza 125
pre-digital generation ages 110
Presser, Jacques 99
'professional help' 100
Prośba o Annę (Czcibor-Piotrowski, Andrzej) 98, 103
PTSD *see* posttraumatic stress disorder
'public squares' 117

Reading Autobiography (Smith & Watson) 7
'real property' 148
Ribbentrop-Molotov treaty 102
Roots (Haley) 10
Rosenstone, Robert A. 8, 11, 13
Rzeczy nienasycone (Czcibor-Piotrowski, Andrzej) 98, 103, 104

Scheppele, Kim Lane 99
Seeger, Peggy 29
self-censorship: in Czcibor-Piotrowski's writing 105; elements of 104; linguistic manifestations of 98; mechanisms of 99–101
Setzler, Frank M. *129*, 130, 137
Silver Action (Lacy) 28
Sisterhood and After 21, 22–3, 28, 29
Smith, Sidonie 2, 7
Smithsonian Institution 134
social media 117
'societal forgetting' 117
Someone Else Can Clean Up This Mess 27
Sorenson, Gerri 27
Soviet Army 102
Soviet invasion of Poland (1939) 101
Soviet Union 102
Spanish rule 146
Spencer, Baldwin 135
Spiegelman, Art 11, 13, 14, 16, 17
Stanley, Liz 26, 42
Stille, Alexander 9–16; *Force of Things, The* 11
Stroińska, Magda 4
Summerfield, Penny 21
Sydney Power-House Museum 4

taboo 124
Taylor, Barbara 32, 37n4; early family life 23; family story 24–5; *Last Asylum, The* (Taylor) 22; memoir 22–3; *Sisterhood and After* 22
'techa' 145

The Man Who Swam into History (Rosenstone) 11
'The Right to Be Forgotten' 116
The Virtue of Forgetting in the Digital Age 117
Thomas, Martin 5, 6n3, *141*
Thompson, Paul 20
Thynne, Lizzie 29–30, 32–4, 36
'tidelands'/'submerged lands' 148
Toker, Leona 98
traumatic event, 'eyewitness' account of 100
traumatic memories 101
tsamorita 146

Ukrainian script 113
United Nations High Commissioner for Human Rights 6n1
unwritten laws of behaviour 100

Valgerist, Emilie 53n2
Van Peenan, Mavis Warner 147
Vera Rado 57; arrival quilting piece *91, 92*; Australia quilt 86, *87*; 'naturalisation' 88; quilting piece 88–9; wartime quilt 75–6, *77, 78*
Vicky van der Ley: Australia quilt 92; childhood quilt *67*, 68
Virgin Mary 147
Voices in Movement 29, 36; *Family Ties: Reframing Memory Exhibition* 29, 31, 35; first iteration of 33; Hughes, Ed 33; at Peltz Gallery 33–5; sound installation 30; Thynne, Lizzie 29–30, 32–4

Walker, Howell 130
Walking since Daybreak (Eksteins) 11, 17
Wamud 127, 128, 140
'Wamud' 124, 125
wartime quilt 57, *59*; Anna Dijkman-Tetteroo 79–81, *80, 81*, 83; Caucasians 72; food drops 82, 83, *84–6*; Frances Larder *69, 70*, 71; Gerada Baremans 78, *79*, 81, *82*; Johanna Binkhorst *68*, 69, *71*, 78, 82, *83*; Nazis' goals for Netherlands 70, 71; queuing for food *83*; Vera Rado 75–6, *77, 78*; Wilhelmina De Brey *73*, 73–5, *74, 76*
Watson, Julia 2, 7
Wet Season of 2012 123
What They Saved (Miller) 8, 11, 14, 52
Widitz, Francis 57
Wilhelmina De Brey, wartime quilt *73*, 73–5, *74, 76*
women's liberation movement (WLM) 21
Women's Liberation Oral History Project 5
Woolf, Virginia 113
Wubarr 126, 130